The **Advanced** Guide to
Microsoft Word 2007
Exam 77-601 Study Guide

another
Computer Mama
Guide

ISTE
This curriculum exceeds the National Education Technology Standards for Secondary Education

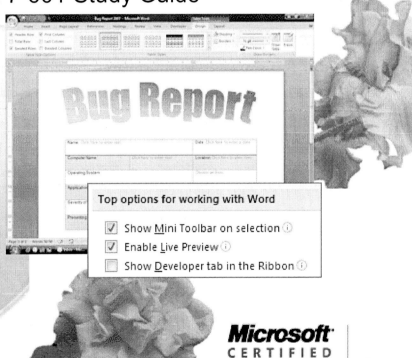

Microsoft
CERTIFIED
Application Specialist

Approved Courseware

© 2008 Comma Productions

The Comma Project

Advanced Guide to Microsoft Word 2007

© 2007-8 Comma Productions
9090 Chilson Road
Brighton, MI 48116
ISBN: 978-0-9818778-7-7

another
**Computer
Mama
Guide**

Trademark and Copyright

Microsoft Word ®, Microsoft Excel ®, Microsoft Access ®, Microsoft Outlook ®, Microsoft PowerPoint ®, Microsoft Windows ® are trademarks or registered trademark of Microsoft Corporation. Adobe Photoshop® is a trademark or register trademark of Adobe Corporation.

Limit of Liability/Disclaimer of Warranty:

The Publisher and Author make no representation or warranties with respect to the accuracy or completeness of the contents of this work and specifically disclaim all warranties including without limitations warranties of fitness for a particular purpose. The advice and strategies contained herein may not be suitable for every situation. The fact that an organization is refereed to in this work as a citation and/or potential source of further information does not mean that the authors or publisher endorses the information that the organization or website may provide or recommendations it may make. Readers should be aware that Internet websites listed in this work may have changed or moved between when this work was written and when it is read.

Neither Comma Productions nor Author shall be liable for any loss of profit or any other commercial damages including but not limited to special, incidental, consequential or other damages.

Comma Products

Comma Project, LLC.

Books Available in this Series:
Beginning Guide to Microsoft® Word 2007

Intermediate Guide to Microsoft® Word 2007

Advanced Guide to Microsoft® Word 2007

About this course

The Advanced Guide to Microsoft Word 2007
Exam 77-601: Using Microsoft® Office Word 2007

Description: *The Advanced Guide to Microsoft Word 2007* demonstrates how to create an on-line form with drop down lists and default text values, create and edit styles, create a Table of Contents, use Headers and Footers, Section Breaks, and the Document Map.

Who will benefit from this course? Students who want to learn editing and layout will enjoy creating interactive forms with drop-down lists. Students can find many advanced functions that will enable them to navigate large documents. Successful students master how to create interactive forms, format Styles and prepare a document for sharing and security.

Audience Description: Who should take this course?
The certification training is available to adult learners online in partnership with major colleges and universities. The audience for this course includes:
- Office workers, managers, entrepreneurs, teachers, and military personnel who want to start using advanced skills immediately
- Job training and professional development (WIA), as well as retired and unemployed people looking to expand their job possibilities

Course Prerequisites: Students who enroll in Microsoft Certified Application Specialist (MCAS) program should have basic computer skills including how to turn on the computer, how to use an Internet browser and how to select commands from a menu or toolbar. Students should complete the *Intermediate Guide to Word 2007* prior to taking this course,

© 2009 Comma Productions

Office Specialist Access Excel Outlook PowerPoint Word Vista

Online Course Requirements
Microsoft Certified Application Specialist (MCAS) certification training

You will need the following Microsoft products already installed on your computer in order to take this course online: Windows Vista Business edition, or Windows XP, For the Microsoft Office 2007 certification course, you MUST have the following software: Word 2007, Excel 2007, Access 2007; Outlook 2007and PowerPoint 2007. Microsoft Office 2007 is NOT the same as Microsoft Office 97-2003.

Hardware requirements for Vista Business:
- IBM-compatible (PC) computer running
- Processor: 1 GHz 32-bit (x86) or 64-bit (x64)
- RAM: 1 GB of system memory **(needs more)**
- Hard Drive: 40 GB with at least 15 GB of available space
- Video: Support for DirectX 9 graphics with WDDM Driver
- 128 MB of graphics memory (needs more)
- Pixel Shader 2.0 in hardware
- 32 bits per pixel
- DVD-ROM drive
- Audio Output

Adobe Acrobat Reader (free version) and a Flash Player.

An online course requires reliable, effect Internet access. If your internet service provider uses only dial-up, a minimum of 56K connection rate is recommended; however, high speed access (Cable or DSL) is preferred. This course cannot be taken with a Macintosh computer

Office Specialist Access Excel Outlook PowerPoint Word Vista

Microsoft Certified Application Specialist

For more information:
Microsoft Business Certification

What is the Microsoft Business Certification Program?

The Microsoft Business Certification Program enables candidates to show that they have something exceptional to offer – proven expertise in Microsoft Office programs. The two certification tracks allow candidates to choose how they want to exhibit their skills, either through validating skills within a specific Microsoft product or taking their knowledge to the next level and combining Microsoft programs to show that they can apply multiple skill sets to complete more complex office tasks. Recognized by businesses and schools around the world, over 3 million certifications have been obtained in over 100 different countries. The Microsoft Business Certification Program is the only Microsoft-approved certification program of its kind.

What is the Microsoft Certified Application Specialist Certification?

The **Microsoft Certified Application Specialist** Certification exams focus on validating specific skill sets within each of the Microsoft® Office system programs. The candidate can choose which exam(s) they want to take according to which skills they want to validate. The available Application Specialist exams include:

Using Microsoft ®Windows Vista™
Using Microsoft® Office Word 2007
Using Microsoft® Office Excel® 2007
Using Microsoft® Office PowerPoint® 2007
Using Microsoft® Office Access 2007
Using Microsoft® Office Outlook® 2007

Please Note: Comma Project, LLC. is independent from Microsoft Corporation, and not affiliated with Microsoft in any manner. While the Complete Computer Guides may be used in assisting individuals to prepare for a Microsoft Business Certification exam, Microsoft, its designated program administrator, and Comma Project, LLC. do not warrant that use of this Complete Computer Guides will ensure passing a Microsoft Business Certification exam.

Exam 77-601: Using Microsoft® Office Word 2007
Microsoft Certified Application Specialist (MCAS) reference topics

Description: The Microsoft Certified Application Specialist program is the only comprehensive, performance-based certification program approved by Microsoft to validate business computer skills using Microsoft Windows Vista® and Microsoft Office® 2007 productivity software: Excel, Word, Power Point, Access, and Outlook.

The Beginning Guide to Microsoft Word 2007 demonstrates how to Cut, copy, and paste, Insert pictures from ClipArt, Undo, Redo, Drag and drop editing, Move text, Insert ClipArt, and Resize pictures, Insert date and time, Insert picture from file, Format font and font size, File Save As a template.

The Intermediate Guide to Microsoft Word 2007 demonstrates how to Use Mail Merge: Create main document, Create data source, Sort records to be merged, Merge the document and data, Create and format tables, Add borders and shading, Merge Cells, Design a Web Page, Create Hyperlinks, Use Design Gallery Live.

The Advanced Guide to Microsoft Word 2007 demonstrates how to Create an on-line form with drop down lists and default text values, Create and edit styles, Create a Table of Contents, Use Headers and Footers, Create Section Breaks, Use the Document Map.

Microsoft Certified Application Specialist (MCAS) objectives for Word 2007

Study Guides
Beginning Word
Intermediate Word
Advanced Word

MCAS Word Word Beginning Word Intermediate **Word Advanced** Advanced Topics Page 1 2

Microsoft Word 2007 Study Guide
Microsoft Certified Application Specialist (MCAS): Microsoft Word 2007 Exam 77-601 Guide

1. Creating and Customizing Documents
Apply Quick Styles to Documents, 44
Apply Themes, 106
Customize Microsoft Word 2007, 131
Customize Microsoft Word 2007: Disable Reading Mode for E-Mail, 136
Customize Microsoft Word 2007: Personalize User Name, Initials, 134
Customize Microsoft Word 2007: Quick Access Toolbar, 70
Customize Themes, 107
Customize Themes: Colors, 108
Customize Themes: Effects, 110
Customize Themes: Fonts, 109
Date and Time: Automatic, 63
Date and Time: Modify, 63
Different First Page, 65
Document Properties: Key Words, 134
Document Properties: Office Button, 134
Format Documents with Themes, 111

Headers and Footers, 62
Headers and Footers: Page No, 62
Indexes: Create, Modify and Update, 73
Indexes: Mark an Entry for Indexing, 74
Insert Blank Pages or Cover Pages, 64
Lay Out Documents: Page Numbers, 67
Research: Change Options, 130
Restore Template Themes, 111
Set Themes as Default, 111
Table of Contents: Add Text, 72
Table of Contents: Create, 69
Table of Contents: Modify Format, 71
Table of Contents: Update, 72

2. Formatting Content
Find and Replace: Replace All, 86
Find and Replace: Replace Text, 85
Find and Replace: Search for Text, 86
Format Characters: Clear Format, 57
Format Paragraphs: Indentation, 37
Format Paragraphs: Spacing, 35
Format Paragraphs: Quotes, 82
Page Breaks: Insert and Delete, 58
Section Breaks, 66
Section Breaks: Delete, 68
Section Breaks: Insert, 66
Section Breaks: Headers, Footers, 67

Styles: Apply Styles, 4
Styles: Change Fonts, 50
Styles: Change Styles, 48
Styles: Create New Style Based on Existing Style, 52
Styles: Create New Style, 53
Styles: Format Body Text, 47
Styles: Format Headings, 45
Styles: Modify Styles, 48
Styles: Reveal Style Formatting, 55
Styles: The Format Painter, 56
Tabs: Clear All Tab Stops, 17
Tabs: Clear One Tab, 17
Tabs: Leaders, 17
Tabs: Set and Clear, 13

MCAS Word   Word Beginning   Word Intermediate   Word Advanced   Advanced Topics Page 1 2

Microsoft Word 2207 Study Guide
Microsoft Certified Application Specialist (MCAS): Microsoft Word 2007 Exam 77-601 Guide

3. Working with Visual Content
Format Text: Drop Caps, 83
Format Text: Pull Quotes, 82
WordArt: Insert and Modify, 102

4. Organizing Content
Convert Tables to Text, 26
Convert Text to Lists, 27
Convert Text to Tables, 18
Lists: Change Bullet Options, 28
Lists: Change Numbering, 31
Lists: Promote and Demote Items, 29
Lists: Sort Items, 33
Quick Parts: Fields, 69
Reference Style: MLA, APA or Chicago Manual, 77
References: Bibliographies, 79
References: Citations, 75
References: Sources, 76
References: Table of Figures, 80
Tables: Change Position and Direction of Text, 22
Tables: Insert and Delete Rows and Columns, 19
Tables: Modify Table Properties, 21
Tables: Perform Calculations, 24
Tables: Sort Contents, 25

5. Reviewing Documents
Comments: Insert and Delete, 129
Find and Go To, 87
Manage Multiple Documents, 124
Track Changes: Accept or Reject, 117
Track Changes: Balloon Options, 120
Track Changes: By Reviewer, 119
Track Changes: Change Options, 120
Track Changes: Display Markup, 118
Track Changes: Enable or Disable, 116
Track Changes: Formatting, 122
Track Changes: Modify Insertions and Deletions, 121
Track Changes: Moves, 123
Track Changes: Reviewer Options, 119
Versions: Combine Revisions for Many Authors, 128
Versions: Compare Documents, 124
Versions: Merge into a New File, 126
Versions: Merge into Existing File, 126

6. Sharing and Securing Content
Digital Signatures, 115
Document Inspector: Options, 113
  Remove Annotations, 113
  Remove Hidden Text, 113
Prepare to Share: Mark as Final, 132
  Restrict Permissions, 99
  Set Editing Restrictions, 99
  Set Formatting Restrictions, 100
  Set Passwords, 100
  Set Passwords, 114

Microsoft Office Beginning Word Intermediate Word Advanced Word

Table of Contents
Beginning Guide to Microsoft® Word 2007

Table for Two
Advanced Table Functions Page 11
Introduction page 12
Working With Tabs page 13
Convert Text to Table page 18
Table Properties page 21
Format the Cell page 22
Calculate the Sum page 24
Sort the Table page 25
Convert Table to Text page 26
Convert Text to List page 27
Format the Bullets page 28
Multilevel List Formats page 29
Define a New List Style page 32
Sort the List page 33
Format the Line Spacing page 35
Indentations page 37

Doing It With Style
Create Styles Page 41
Introduction to Styles page 42
Create and Edit Styles page 45
Modify the Style Properties page 50
Insert A Page Break page 58
Use the Document Map page 60
Headers and Footers page 61
Add a Cover Page page 64
Different First Page page 65
Use Section Breaks page 66
Create a Table of Contents page 69
Create an Index page 73
Create a Citation page 75
Create a Table of Figures page 80
Format Quoted Material page 81
Drop Caps page 82
Find, Replace and Go page 85

Who Done It?
Create an on-line form Page 89
Create and Modify Forms page 90
Design with a Table page 91
Find the Developer Toolbar page 94
Make Text Controls page 95
Create a Date Picker page 96
Make a Combo Box page 97
Protect the Document page 99
Format the Form page 102
Format for Print page 104
Apply Themes page 106
Prepare to Share page 112
Track Changes page 116
Compare Documents page 124
Research Options page 130
Customize Word 2007 page 145

Microsoft Office Beginning Word Intermediate Word Advanced Word

Table of Contents, continued
Beginning Guide to Microsoft® Word 2007

2. Downloads
Samples

2. Assessment
Performance
Multiple Choice

Resources
Lesson Plan
Exercises
Menu Map

Objectives
Exploring Strategies
Formatting type and creating subtotals are repetitive tasks. A document or spreadsheet can contain hundreds of words or cells that need to be modified.

With computers, you can tell the computer how to do it right, once, then let the machine track all of the updates.

Questions: Formatting Styles make **dependencies** in Word.

You can use dependencies to make global updates. Does Excel have a method for analyzing data that exploits dependencies?

Page 1 2 3 4 5 6 7 8 9 10 11 12 13 14 More

Charlotte's Website
Table for Two

Click Here To Get Started
Sample Files

Advanced Word
Lesson Objectives: This lesson shows how to organize and format content as Tabs, Tables and Lists. In this lesson you will:

Identify options for working with Tabs page 3

Learn the steps to Convert Text to Table page 8

Investigate the Table Properties page 11

Practice how to format a Cell page 12

Use the Layout Ribbon to calculate the Sum page 14

Show how to Convert Table to Text page 16

Show how to Convert Text to List page 17

Use the Paragraph Group to Format Bullets page 18

Use the Paragraph Group for Multilevel Lists page 19

Investigate Line Spacing options page 25

Use the Paragraph Group to edit Indentations 27

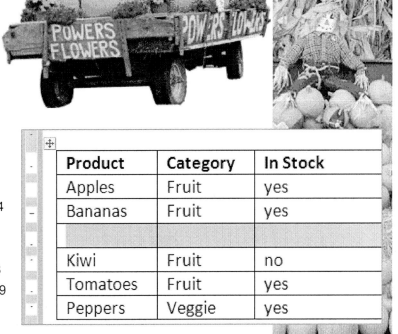

Product	Category	In Stock
Apples	Fruit	yes
Bananas	Fruit	yes
Kiwi	Fruit	no
Tomatoes	Fruit	yes
Peppers	Veggie	yes

Word: Table for Two Page 1 2 3 4 5 6 7 8 9 10 11 12 13 14 More

A Table is a Table is a Table

A table is a fundamental way for organizing information into rows and columns. Creating tables is one of the unique functions of a computer. Microsoft Word is an excellent tool for designing, formatting and using tables and lists. Many of the demonstrations in this lesson echo the same commands and options that you can find in Microsoft Excel and Microsoft Access. **Tables are interchangeable**: you can use a table in Excel for a Mail Merge or link that information to Access.

This lesson begins with ancient technology: **Tabs**. Typewriters used Tabs to create lists and columns. Very quickly you will see the limitations of using Tabs. All digital data can be recycled, so the lesson demonstrates how to **Convert the Text** (and all of the Tabs) to a Table.

The key to working with Tables is to watch the commands on the Ribbons.

Word: Table for Two Page 1 2 3 4 5 6 7 8 9 10 11 12 13 14 More

Home -> Paragraph ->Show/Hide

Working with Tabs

Try This: Enter the Sample Text
Type the text that you see on this page. Add one Tab between each item: meaning type the name of the fruit and click on the tab key on your keyboard. When you complete a Row, click on the Enter key to start the next Row.

Try This, Too: Proof Your Typing
Go to **Home ->Paragraph** and select the **Show/Hide** button to reveal the Tab stops. Show/Hide looks like a backwards "P."

Please note: the Tabs will be uneven. They will not line up like a column, yet.
Keep going...

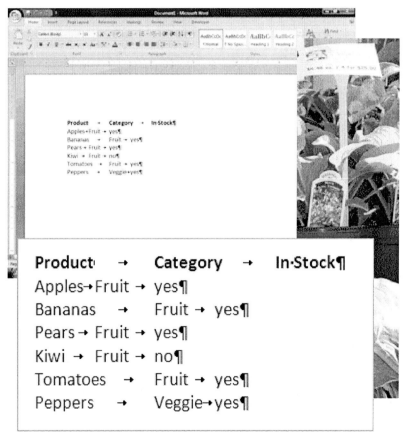

Product → Category → In·Stock¶
Apples→Fruit → yes¶
Bananas → Fruit → yes¶
Pears→ Fruit → yes¶
Kiwi → Fruit → no¶
Tomatoes → Fruit → yes¶
Peppers → Veggie→yes¶

Word: Table for Two Page 1 2 3 **4** 5 6 7 8 9 10 11 12 13 14 More

Home -> Paragraph ->Show/Hide

Tab Stops
Try This: Find the Tab Stops
The Tab Stops are on the Ruler on the far left side. Click on the "L" and review the options, please.

This Tab aligns the text Left.

This Tab aligns the text Center.

This Tab aligns the text Left.

This Tab aligns the text Right.

The Bar Tab creates a line.

This Tab aligns the indent on the first line of type in a paragraph.

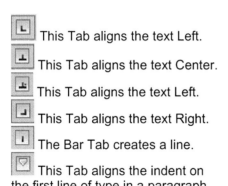

Microsoft Word 2007 Exam 77-601 Topic: 2. Formatting Content
2.1. Format text and paragraphs
2.1.5. Set and clear tabs

Word: Table for Two Page 1 2 3 4 **5** 6 7 8 9 10 11 12 13 14 More

Home -> Paragraph ->Show/Hide

Add the Tabs
Try This: Create the Tab Stops
Select all of the sample text.
Select the Left Tab Stop.
Go to the top ruler and click on the 1" mark and the 2" mark.

Each time you click on the ruler you will place another Left Tab stop. If you add too many Tabs, you can drag them off the ruler to get rid of them.

What Do You See? The Tab Stops will align the sample text into columns.

Can you adjust the Tab stops by dragging them left or right on the ruler?

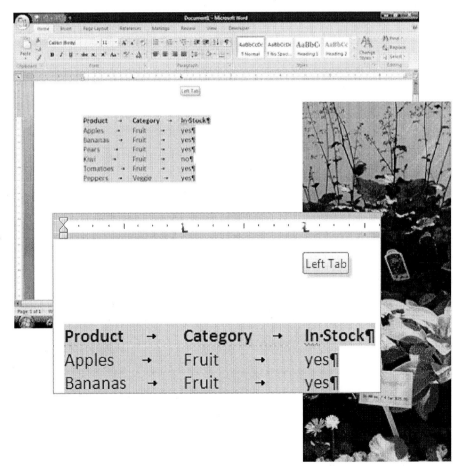

Microsoft Word 2007 Exam 77-601 Topic: 2. Formatting Content
2.1. Format text and paragraphs
2.1.5. Set and clear tabs

Word: Table for Two Page 1 2 3 4 5 6 7 8 9 10 11 12 13 14 More

Home -> Paragraph ->Paragraph

Modify the Tabs

Tab properties are bundled with the **Paragraph** tools. Paragraph is on the **Home** Ribbon.

Click on the small arrow on the bottom right side.

What Do You See? When you click on the arrow a new window will open. The Paragraph properties include alignment, indentation and line spacing.

Look at the bottom of this window for the **Tabs** button.

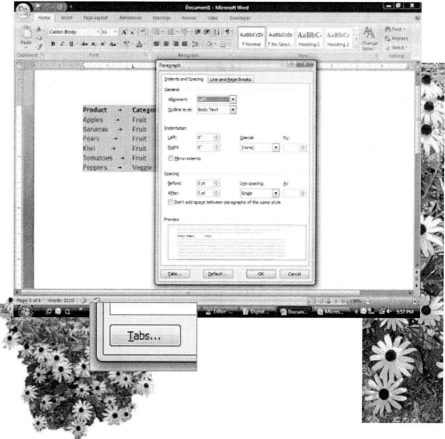

Microsoft Word 2007 Exam 77-601 Topic: 2. Formatting Content
2.1. Format text and paragraphs
2.1.5. Set and clear tabs

Word: Table for Two Page 1 2 3 4 5 6 7 8 9 10 11 12 13 14 More

Home -> Paragraph ->Paragraph

Tab Stops
Each Tab stop is identified by it's position. When you select a Tab stop you can edit the position as well as the Alignment and the Leader.

Try This: Change the Leader
Select the first Tab stop and choose the #4 Leader. Click on Set.

Select the second Tab stop and choose the #4 Leader. Click on **OK**.

Memo to Self: You can use this window to **Clear** one or **Clear All** Tab stops.

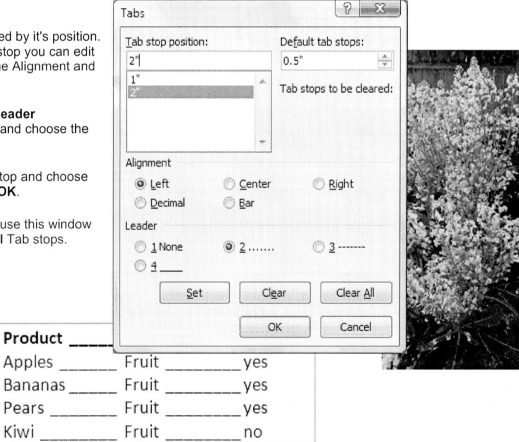

Microsoft Word 2007 Exam 77-601 Topic: 2. Formatting Content
2.1. Format text and paragraphs
2.1.5. Set and clear tabs: Tabs with leaders

Word: Table for Two Page 1 2 3 4 5 6 7 8 9 10 11 12 13 14 More

Insert ->Table ->Convert Text to Table

Convert Text to Table

This is about as far as you can go with Tabs. **Tabs** are not an effective method for creating rows and columns. It is very difficult to add another column of data to a tabbed list. A **Table** offers a better way to organize information.

Try This: Convert the Text to a Table
Select all of the tabbed text.
Go to **Insert ->Table**.
Select: **Convert Text to Table**.

You will be prompted to confirm how many columns there are in this table.

Memo to Self: This sample list has 3 columns. If you see more than 3 columns, then some of your rows have more than one tab stop between the items. You can use the **Show/Hide** function to see each Tab Stop as an arrow and proof your work.

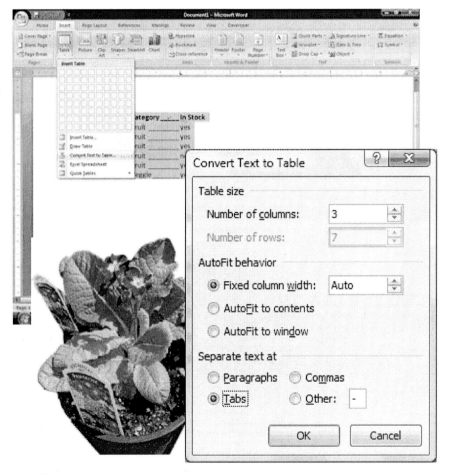

Microsoft Word 2007 Exam 77-601 Topic: 4. Organizing Content
4.2. Use tables and lists to organize content
4.2.1. Create tables and lists: Convert text to tables

Word: Table for Two Page 1 2 3 4 5 6 7 8 9 10 11 12 13 14 More

Table Tools -> Layout -> Rows and Columns

Insert a New Column
It is very easy to modify a table, compared to a tabbed list.

Try It: Insert A Column
Select Column C.
Go to **Table Tools ->Layout**.
Select: **Insert Right**

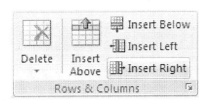

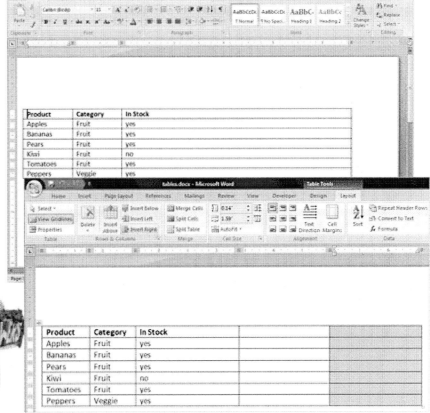

Word: Table for Two Page 1 2 3 4 5 6 7 8 9 **10** 11 12 13 14 More

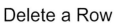

Table Tools -> Layout -> Rows and Columns

Delete a Row

If you select a row and delete it, will the row be deleted, or just the text?

Try This: Delete a Row
Select Row 4: Pears.
Press **Delete** on your keyboard.

What Do You See? The data was deleted, but the row is still there.

Try This Instead: Delete the Row
Select Row 4.
Go to **Table Tools -> Layout**.
Click on **Delete -> Rows**.

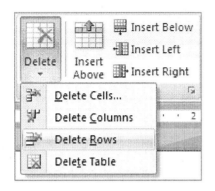

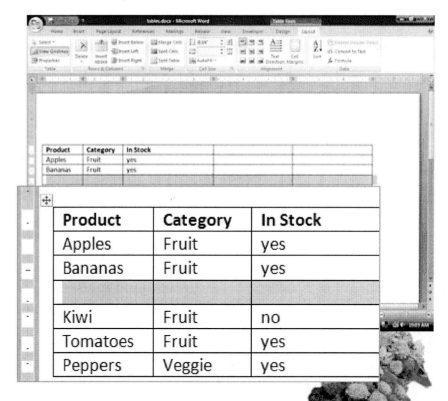

You can delete the two blank columns you added, too.

Microsoft Word 2007 Exam 77-601 Topic: 4. Organizing Content
4.3. Modify tables
4.3.2. Modify table properties and options: Insert and delete rows and columns

Word: Table for Two Page 1 2 3 4 5 6 7 8 9 10 **11** 12 13 14 More

Table Tools -> Layout -> Properties

Table Properties

Tables have Properties. The Properties include options for formatting the Table, Row, Column and Cell.

Review the Table Properties
Select the Table.
Go to **Table Tools ->Layout**
Select: **Properties**.

What Do You See? The Table is an object just like a picture or graphic. A Table can be **Aligned** Left, Center, or Right on a page.

You can also use **Text Wrapping** to edit how text wraps around your table.

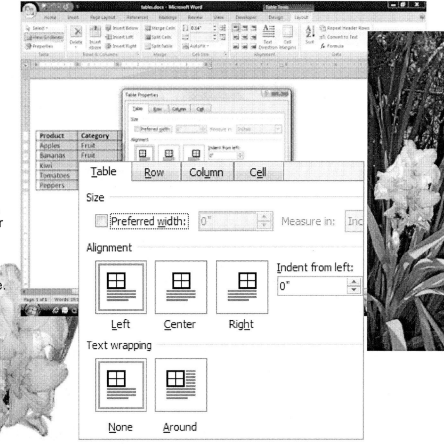

Microsoft Word 2007 Exam 77-601 Topic: 4. Organizing Content
4.3. Modify tables
4.3.2. Modify table properties and options

Word: Table for Two Page 1 2 3 4 5 6 7 8 9 10 11 **12** 13 14 More

Table Tools -> Layout -> Properties

Format the Cell
Try This: Edit the Cell Properties
Select the Table.
Go to **Table Tools ->Layout**
Select: **Properties.**

The Cell options include controlling the **Vertical alignment**. You can choose whether the text in the table is placed at the Top, Center, or Bottom of the Cells.

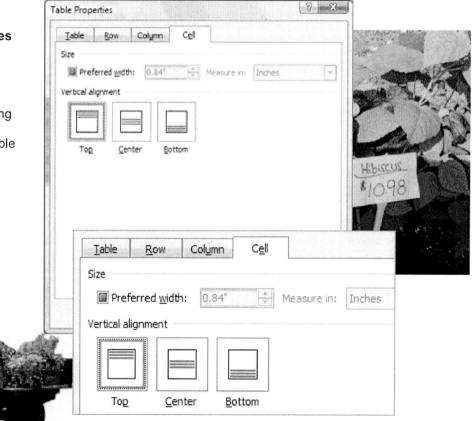

Microsoft Word 2007 Exam 77-601 Topic: 4. Organizing Content
4.3. Modify tables
4.3.5. Change the position and direction of cell contents

Word: Table for Two Page 1 2 3 4 5 6 7 8 9 10 11 12 13 14 More

Table Tools -> Layout -> Text Direction

Format the Alignment

There are additional Table Tools for the Text in the Cells. These tools can be found on the Layout Ribbon.

Try This: Modify the Direction
Select the first row in this table.
Go to **Table Tools -> Layout**.
Select **Text Direction**.

What Do You See? This is a simple command. It does not have a drop down list of options. There are a 3 different directions that you can select as you click again and again.

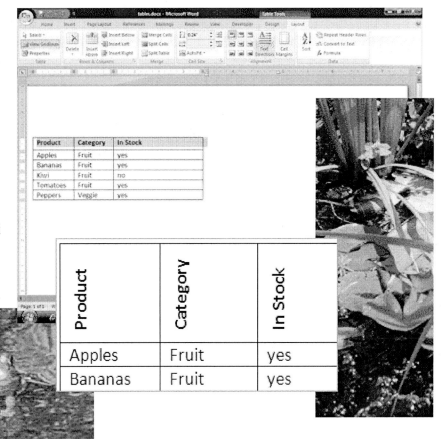

Microsoft Word 2007 Exam 77-601 Topic: 4. Organizing Content
4.3. Modify tables
4.3.5. Change the position and direction of cell contents

Word: Table for Two Page 1 2 3 4 5 6 7 8 9 10 11 12 13 14 More

Table Tools -> Layout -> Formula

Calculate the Sum
You can calculate the Sum in a table, although it is not as simple as working with a formula in Microsoft Excel.

Try This: Calculate the Sum
Enter the sample data: please edit the last column in your sample table and type in the quantity (in numbers) for your fruit and veggie sales.

Insert a new, blank row at the bottom of your table. Place your cursor in the last column. Enter the formula here.

Go to **Table Tools -> Layout.**
Select : **Formula.**

What Do You See? Microsoft Word suggests the formula:
=SUM(ABOVE).

You can choose which number format is appropriate: general, accounting, and percentage.

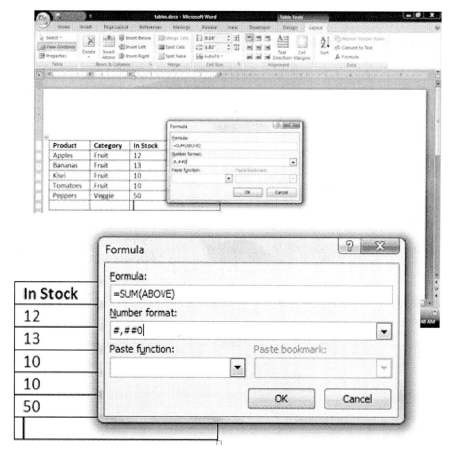

Word: Table for Two (Page 1) 14 15 16 17 18 19 20 21 22 23 24 25 26 27 28 29

Table Tools -> Layout -> Sort

Sort the Table
Information in a Microsoft Word table can be **Sorted**. Here are the options:

Try This: Sort the Table
Select the entire table.
Go to **Table Tools ->Layout**.
Click on **Sort**.

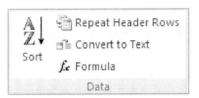

What Do You See? You can choose which field you want to Sort. In the example on this page, the table will be Sorted by the Products.

Look At The Bottom of the Window: Does it say, **My list has a Header Row**? That is the first row with the labels.

Microsoft Word 2007 Exam 77-601 Topic: 4. Organizing Content
4.2. Use tables and lists to organize content
4.2.2. Sort content: Sort table contents

Word: Table for Two (Page 1) 14 15 16 17 18 19 20 21 22 23 24 25 26 27 28 29

Table Tools -> Layout -> Convert to Text

Convert Table to Text
There and back again: here are the steps to convert a Table back to Text.

Try This: Convert Table to Text
Select the Table.
Go to **Table Tools -> Layout**.
Select: **Convert to Text**.

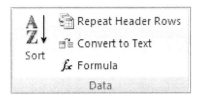

What Do You See? You will be prompted to select how you would prefer to convert the columns. You can choose Paragraphs, Tabs, Commas, or Other symbols to separate your text.

In this example, select Commas.

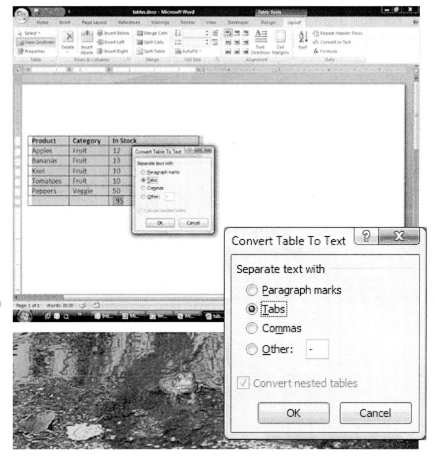

Microsoft Word 2007 Exam 77-601 Topic: 4. Organizing Content
4.2. Use tables and lists to organize content
4.2.2. Sort content: Sort table contents

Word: Table for Two (Page 1) 14 15 16 17 18 19 20 21 22 23 24 25 26 27 28 29

Home -> Paragraph -> Bullets

Convert Text to List

The bulleted or numbered list is another method for organizing and presenting your data.

Try This: Convert Text to List.
Select the sample text.
Go to **Home -> Paragraph**.
Click on **Bullets**.

What Do You See? You can use the down-arrow to select a different bullet from the library. The selected text is indented. There are bullets at the beginning of each line of type.

Keep going....

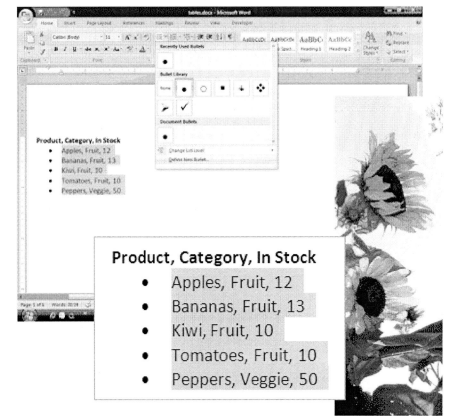

Word: Table for Two (Page 1) 14 15 16 17 18 19 20 21 22 23 24 25 26 27 28 29

Home -> Paragraph -> Paragraph

Format the Bullets
You can choose a different **Symbol** or **Picture** for your bullets.

Try This: Define New Bullet
Select the sample text.
Go to **Home -> Paragraph**.
Click on the **Bullets** down-arrow.
Go to the bottom of the list and select **Define New Bullet**.

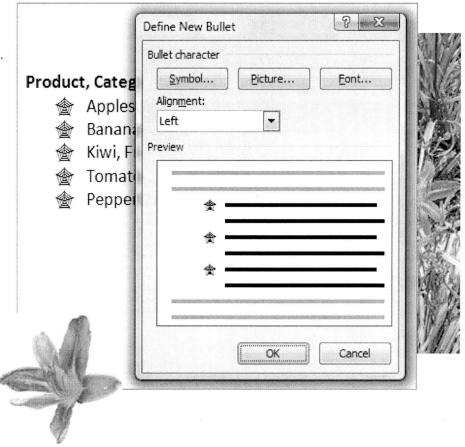

When you click on **Symbol**, you can review the characters in the WingDing type.

The **Picture** button offers little graphics that match various themes in Microsoft Office. You can also select your own picture. Keep it small and simple!

Word: Table for Two (Page 1) 14 15 16 17 18 19 20 21 22 23 24 25 26 27 28 29

Home -> Paragraph -> Multilevel List

Multilevel List Formats
A **Multilevel List** organizes your information into topics and subtopics. These outlines can be formatted with several built-in Styles.

Try It: Create a Multilevel List
Select the sample text.
Go to **Home -> Paragraph**.
Select: **Multilevel List**.

What Do You See? The List Library offers several outline formats. Please select the first Style.

1)____
 a)____
  i) ___

Keep going....

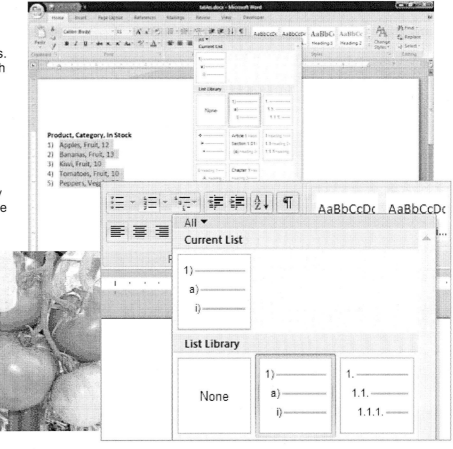

Microsoft Word 2007 Exam 77-601 Topic: 4. Organizing Content
4.2. Use tables and lists to organize content
4.2.3. Modify list formats: Promote and demote list items

Word: Table for Two (Page 1) 14 15 16 17 18 19 20 21 22 23 24 25 26 27 28 29

Home -> Paragraph -> Demote

Multilevel List Options

A **Multilevel List** format is dynamic and changes as you edit your text. These screenshots demonstrate how to work with the levels,

Try It: Promote and Demote Items
Place your cursor at the end of the line in the sample text.

Press **Enter**: create a new line.

Go to **Home ->Paragraph**
Select **Demote** to indent the cursor.
Type: Roma

Can you add another line? What happens if you **Promote** the text?

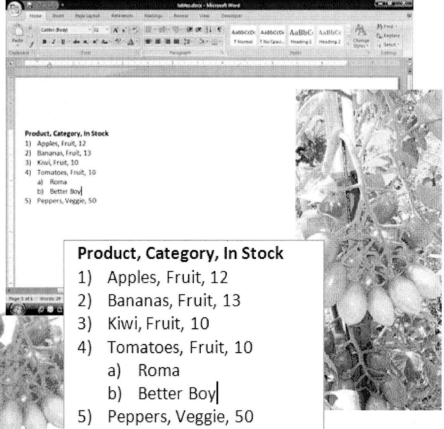

Product, Category, In Stock
1) Apples, Fruit, 12
2) Bananas, Fruit, 13
3) Kiwi, Fruit, 10
4) Tomatoes, Fruit, 10
   a) Roma
   b) Better Boy
5) Peppers, Veggie, 50

Microsoft Word 2007 Exam 77-601 Topic: 4. Organizing Content
4.2. Use tables and lists to organize content
4.2.3. Modify list formats: Promote and demote list items

Word: Table for Two (Page 1) 14 15 16 17 18 19 20 21 22 23 24 25 26 27 28 29

Home -> Paragraph -> Define New List Style

More Number Options
Some professions have very specific formatting. Legal documents are one example. Here are the steps you can take to define your own Style.

Try This: Define New List Style
Select the sample list.
Go to **Home ->Paragraph.**
Select Define New List Style.

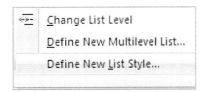

Keep going, please...

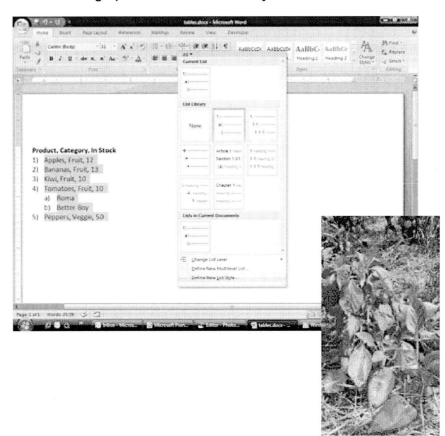

Word: Table for Two (Page 1) 14 15 16 17 18 19 20 21 22 23 24 25 26 27 28 29

Home -> Paragraph -> Define New List Style

Define a New List Style

A new list style is defined one level at a time. You can edit the font, size, color as well as the indentation.

You can select the numbering format for each level. The options include Roman numerals, letters, numbers as well as bullets.

The new **List Style** can be saved with the current document you are working on. You can also choose to add the new List Style to this template. When you add the Style to the template, then it will be available in all new documents which are created with this template.

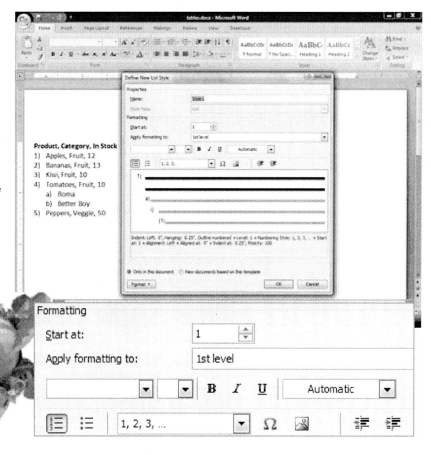

Microsoft Word 2007 Exam 77-601 Topic: 4. Organizing Content
4.2. Use tables and lists to organize content
4.2.3. Modify list formats: Change numbering options

Word: Table for Two (Page 1) 14 15 16 17 18 19 20 21 22 23 24 25 26 27 28 29

Home -> Paragraph ->Sort

Sort the List
Try This: Sort the List
Select the sample list.
Go to **Home ->Paragraph ->Sort**.

What Do You See? The **Sort** options are similar to the ones in Microsoft Excel. This is **Text**, so you can **Sort by Paragraphs**.

Depending on what you selected, your list may or may not have a Header Row, the first row with the labels.

Keep going...

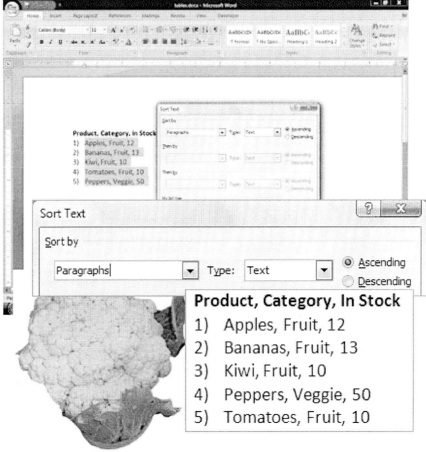

Microsoft Word 2007 Exam 77-601 Topic: 4. Organizing Content
4.2. Use tables and lists to organize content
4.2.2. Sort content: Sort list items

Word: Table for Two (Page 1) 14 15 16 17 18 19 20 21 22 23 24 25 26 27 28 29

Home -> Paragraph ->Sort

Sort by Paragraphs

A simple list sorts the way you expect: alphabetical, A-Z or vice versa. What happens if you sort a list that has sub items?

Try This: Change the Paragraph
Enter the sample text:

Product, Category, In Stock
1) Apples, Fruit, 12
2) Bananas, Fruit, 13
3) Kiwi, Fruit, 10
4) Tomatoes, Fruit, 10
  a) Roma
  b) Better Boy
5) Peppers, Veggie, 50

Use the bullets and numbering options to format this list as an outline.

Select the sample list.
Go to Home ->Paragraph ->Sort.

The results may come out scrambled.

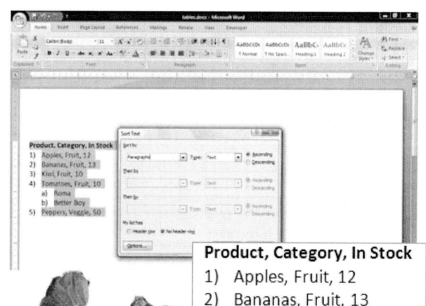

Product, Category, In Stock
1) Apples, Fruit, 12
2) Bananas, Fruit, 13
  a) Better Boy
3) Kiwi, Fruit, 10
4) Peppers, Veggie, 50
  a) Roma
5) Tomatoes, Fruit, 10

Microsoft Word 2007 Exam 77-601 Topic: 4. Organizing Content
4.2. Use tables and lists to organize content
4.2.2. Sort content: Sort list items

Word: Table for Two (Page 1) 14 15 16 17 18 19 20 21 22 23 24 **25** 26 27 28 29

Home -> Paragraph ->Line Spacing

Format the Line Spacing

A bulleted list can be easier to read if you add more space between the lines. The technical term for extra space between the lines is called leading. Here are some options you can try.

Try This: Change the Line Spacing
Select the sample list.
Go to Home ->Paragraph.
Select: **Line Spacing.**

What Do You See? You can increase the amount of space between the rows of text from 1.0 to 3.0.

You can also increase the space between the paragraphs, or bulleted lists in this example, by selecting **Add Space Before Paragraph**.

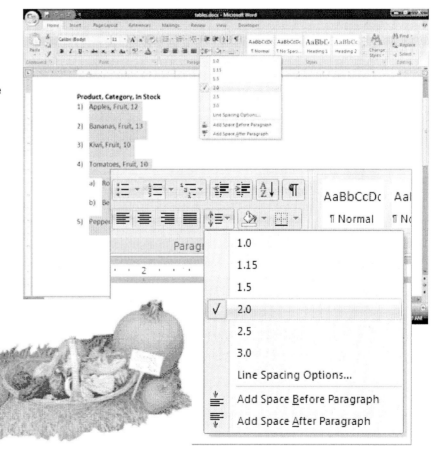

Word: Table for Two (Page 1) 14 15 16 17 18 19 20 21 22 23 24 25 **26** 27 28 29

Home -> Paragraph ->Line Spacing Options

Paragraph Spacing

Try This: Line Spacing
Select the sample list.
Go to **Home ->Paragraph**.
Select: **Line Spacing Options**.

What Do You See? Go to the page for Indents and Spacing. Consider this: There are two parts to formatting paragraphs. You can adjust the leading between the lines of text. That's the **line spacing**. In the example on this page, the line spacing is double, a term that came from 20th century typewriters. You can also adjust the space between paragraphs. This is the **paragraph spacing**. In this screen shot, the Spacing Before and After the paragraph is zero.

Paragraph Spacing Adding 6 pts of extra white **space** (the technical term is leading) **after a paragraph** or bulleted list makes it easier to find and read.

Microsoft Word 2007 Exam 77-601 Topic: 2. Formatting Content
2.1. Format text and paragraphs
2.1.4. Format paragraphs: Change paragraph spacing

Word: Table for Two (Page 1) 14 15 16 17 18 19 20 21 22 23 24 25 26 **27** 28 29

Home -> Paragraph ->Indent

Indentations

A bulleted or outlined list has a special set of tab stops on the top ruler.

The **First Line Indent** adjusts the bullet or number. In this example, the First Line Indent for 1) Apples has been increased by dragging the upper tab stop to the right.

The **Left Indent,** the bottom two tabs on the ruler, adjusts the spacing between the number and the text.

Keep going...

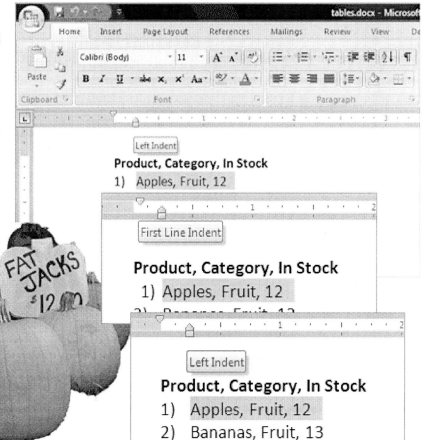

Microsoft Word 2007 Exam 77-601 Topic: 2. Formatting Content
2.1. Format text and paragraphs
2.1.4. Format paragraphs: Change indentation

Word: Table for Two (Page 1) 14 15 16 17 18 19 20 21 22 23 24 25 26 27 28 29

Home -> Paragraph ->Paragraph

Indentations

You can make simple adjustments with the indent buttons in the **Paragraph** group: increase and decrease.

There are additional options you can edit by clicking on the the small arrow in the lower right corner.

Try It: Edit the Indentation
Go to **Home -> Paragraph ->Paragraph**.

What Do You See? The **Left** indent aligns the bullets or numbers in the list.

The **Hanging** indent aligns the text .25" away from the bullets.

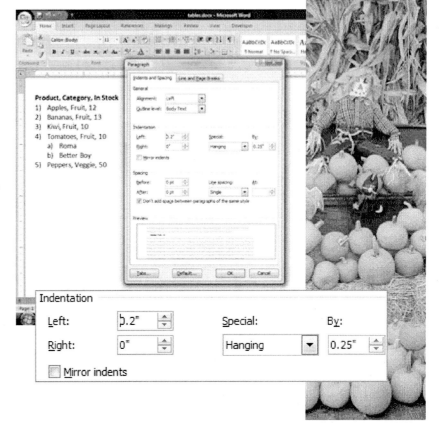

Word: Table for Two (Page 1) 14 15 16 17 18 19 20 21 22 23 24 25 26 27 28 **29**

DONE

Table for Two

This lesson introduced the benefits and limitations of formatting Tab Stops.

All data can be recycled, so the lesson demonstrated how to convert the text into Tables as well as Text to Lists.

Well, you done good.
You get the cookie.

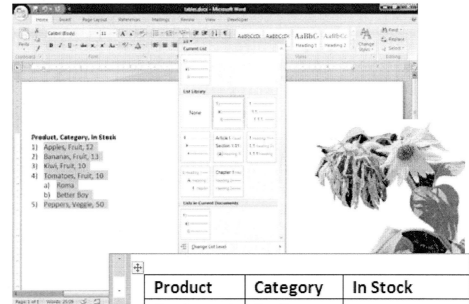

Product	Category	In Stock
Apples	Fruit	yes
Bananas	Fruit	yes
Kiwi	Fruit	no
Tomatoes	Fruit	yes
Peppers	Veggie	yes

Test Yourself

1. Which of the following is a way to set Tab Stops?
a. Click the ruler to insert a tab stop at that location
b. Go to Home->Paragraph ->More Options (little arrow) ->Tabs…
  Tip: Advanced Word, page 15, 16, 17

2. Covert Text to Table will insert a table and put the all selected text into the first cell of the table.
a. TRUE
b. FALSE
  Tip: Advanced Word, page 18,

3. Which of the following is a way to delete a Row from a Table?
a. Select the row, press Delete
b. Select the row, go to Table Tools-> Layout-> Delete->Rows
  Tip: Advanced Word, page 20

4. Tables can be aligned right, left or center?
a. TRUE
b. FALSE
  Tip: Advanced Word, page 21

5. Tables can have text wrapping.
a. TRUE
b. FALSE
  Tip: Advanced Word, page 21

6. Text in a cell can only be at the top of cell.
a. TRUE
b. FALSE
  Tip: Advanced Word, page 22

7. To change the Text Direction, select the cells, go to Table Too->Layout, and Format Cells.
a. TRUE
b. FALSE
  Tip: Advanced Word, page 23

8. To use a formula in a Word Table, go to Table Tools-> Layout -> Formula
a. TRUE
b. FALSE
  Tip: Advanced Word, page 24

9. To Convert Text to list, go to Home -> Paragraph -> Bullets
a. TRUE
b. FALSE
  Tip: Advanced Word, page 27

10. What does List Style include? (Select all correct answers.)
a. Font formatting
b. Indentation
c. Numbering format for each level
  Tip: Advanced Word, page 32

11 Indentations can be changed by going to Home -> Paragraph -> Paragraph Options (blue arrow)
a. TRUE
b. FALSE
  Tip: Advanced Word, page 38

Page 1 2 3 4 5 6 7 8 9 10 11 12 13 14 15 16 17 18 19 20 21 22 More

Get Smart
Doing It With Style

Click Here to Get Started
Sample Files

Advanced Word

Lesson Objectives: Learn how to format text with Styles. This lesson will demonstrate how to use Styles to format text and create document navigation including a Table of Contents, Index and Document Map. In this lesson you will:

Learn how to create, apply and update Styles page 2

Learn how to control pagination with a Page Break page 18

Practice navigating with the Document Map page 20

Learn how to insert Headers and Footers page 21

Practice the steps needed to insert a cover page Page 24

Learn how to use Section Breaks page 26

Identify the steps needed to create a Table of Contents page 29

Use the Reference Ribbon to create References page 33

Identify Editing options for Find, Replace and Go page 45

Word: Styles Page 1 2 3 4 5 6 7 8 9 10 11 12 13 14 15 16 17 18 19 20 21 22 More

Doing It With Style

The default font in Microsoft Word is a good type face, but it gets old. What you want is a little more **STYLE**. Styles are easy to use and can be changed very quickly. Using Styles also makes creating a Document Map or Table of Contents a snap.

Where does Word keep these settings? Is there a way to set up Word so that each new document uses your typeface, size and color? Let's find out. **Start** the Program Microsoft **Word**.

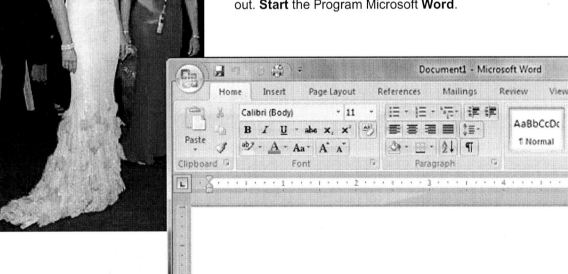

Word: Styles Page 1 2 3 4 5 6 7 8 9 10 11 12 13 14 15 16 17 18 19 20 21 22 More

Office -> Open

Working with Large Documents

This project uses a sample document. Please download a copy of the **sample text** here: 2007 Sample Deployment Text.

This lesson teaches you how to automate text formatting with **Styles**. The Styles, in turn, can be used to create a **Document Map** and a **Table of Contents**.

1. Open the Sample File
Go to **Office-> Open**
Browse to the **Documents** folder
Select: Windows Deployment Text

Word: Styles Page 1 2 3 **4** 5 6 7 8 9 10 11 12 13 14 15 16 17 18 19 20 21 22 More

Home -> Styles

Format the Type

All of the text in the sample Windows Deployment document you opened is the same size, font and weight.

Usually, we just select the text and format the Font with the options on the **Home** Ribbon.

Look past the **Clipboard, Font** and **Paragraph** groups. The **Style** library includes Headings, Titles and Normal. The new Office 2007 file format, *.docx, includes several web styles: Strong, Emphasis, and Intense.

Keep going...

Microsoft Word 2007 Exam 77-601 Topic: 1. Creating and Customizing Documents
1.1.2. Apply Quick Styles to documents

Word: Styles Page 1 2 3 4 **5** 6 7 8 9 10 **11** 12 13 14 15 16 **17** 18 19 20 21 22 More

Home -> Styles

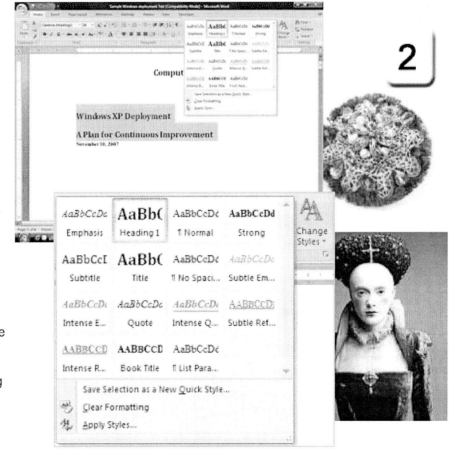

She's Just My Style
2. Format the Styles
Try This: Highlight the text
Select: Computers Are Us.
Select **Title** from the **Styles**.

Select the next two lines of type
Select **Heading 1** from the **Styles**.

Try This, Too: Highlight the date
Select **Strong** from the **Style** library
Strong is a term that is also used in formatting HTML text for web pages. It means bold.

In a 6-8 page document, there may be dozens of headlines. Let's see: if you change the type font, size, and color for 20 headlines that would require 60 actions. What are the chances that you will make a mistake and forget one of the attributes?

When you use **Styles**, the formatting is consistent and efficient.

Microsoft Word 2007 Exam 77-601 Topic: 2. Formatting Content
2.1. Format text and paragraphs
2.1.1. Apply styles: Format headings

Word: Styles Page 1 2 3 4 5 6 7 8 9 10 11 12 13 14 15 16 17 18 19 20 21 22 More

Home -> Styles and Formatting

That's Not My Style
3. Apply the Styles
There are some important benefits that come with using Styles. This lesson makes more sense if you do the homework and format the Styles.

Format these headlines as **Heading 1**:
Business Organization
Issues of Concern
Management Summary
Project Summary
Phase One: Proposed Transition
Phase Two: Migration
Phase Three: The Plan for Continuous Improvement

Each of these paragraphs has a sub heading, too. For example, the minor headings on the first page are:
Security
Fault Tolerance
Recovery

Format the sub headings as **Strong.**

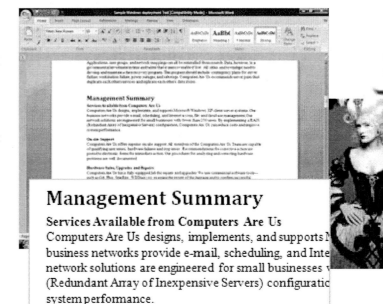

Microsoft Word 2007 Exam 77-601 Topic: 2. Formatting Content
2.1. Format text and paragraphs
2.1.1. Apply styles: Format headings

Word: Styles Page 1 2 3 4 5 6 7 8 9 10 11 12 13 14 15 16 17 18 19 20 21 22 More

Home -> Styles and Formatting

Format Body Text

Ordinary words that aren't headings can be formatted as **body text**. In this example, the body text is highlighted and formatted with the **Normal** Style.

Times New Roman 11 pt was the default typeface and size for the Normal Style Microsoft Word for 10 years!

The default typeface in Microsoft Word 2007 is Calibri.

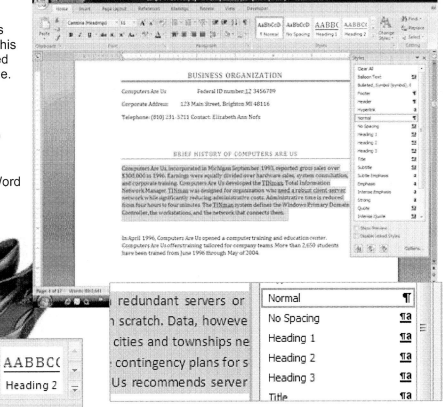

Microsoft Word 2007 Exam 77-601 Topic: 2. Formatting Content
2.1. Format text and paragraphs
2.1.1. Apply styles: Format body text

Word: Styles Page 1 2 3 4 5 6 7 8 9 10 11 12 13 14 15 16 17 18 19 20 21 22 More

Home -> Styles -> Change Styles

Change the Style
Did you notice that selecting a **Style** changed the type size as well as the line spacing and font? Microsoft Word has a Style Gallery that automatically formats your document.

4. Change Styles
Click on **Change Styles**
Go to **Style Set**
Play with the formatting options.

When you select one of the Style Sets, all of the Heading 1, Strong, and Normal text will be reformatted **automatically**.

Microsoft Word 2007 Exam 77-601 Topic: 2. Formatting Content
2.1. Format text and paragraphs
2.1.1. Apply styles: Change from one style to another

Word: Styles Page 1 2 3 4 5 6 7 8 9 10 11 12 13 14 15 16 17 18 19 20 21 22 More

Home -> Styles -> Options

Edit the Style

You can create your own **Styles** to fit your business image. In the Style options, you can change the font, the size, the color and the line spacing.

Try it: Edit the Style
Go to the **Home** Ribbon
Find the **Styles** Group
Look in the bottom right-hand corner for the **Options** arrow.

You should see the **Styles** Window. When you select the words Windows XP Deployment on page 1, you will see Heading 1 highlighted on the list.

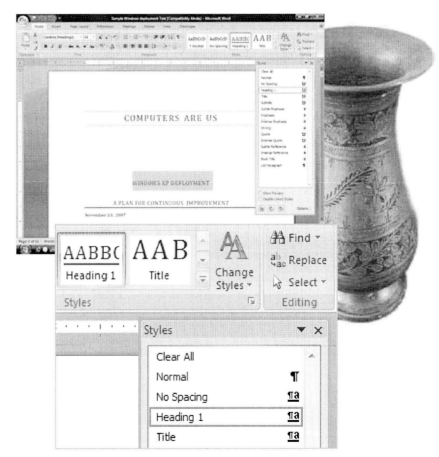

Microsoft Word 2007 Exam 77-601 Topic: 2. Formatting Content
2.1. Format text and paragraphs
2.1.2. Create and modify styles: Change fonts

Word: Styles Page 1 2 3 4 5 6 7 8 9 <u>10</u> 11 12 13 14 15 16 17 18 <u>19</u> 20 21 22 More

Home -> Styles -> Options -> Modify

Modify the Style
Before you **Modify the Style**, you can **Select All** of the text that was formatted with this Style. This step is a good example of Microsoft Office automation: you can change something once and have all the places that refer to it update at the same time.

Try it: Modify the Style
Select the text: Windows XP Deployment on page 1.

Go to **Home->Styles->Options**
Choose **Heading 1** from the Styles.

Select **All Instances** of Heading 1.

Click on **Modify**.

Microsoft Word 2007 Exam 77-601 Topic: 2. Formatting Content
2.1. Format text and paragraphs
2.1.2. Create and modify styles

Word: Styles Page 1 2 3 4 5 6 7 8 9 10 **11** 12 13 14 15 16 17 18 19 20 21 22 More

Home -> Styles -> Options -> Modify

Format the Style
You can use the options here to format the font, size, color and alignment. You can also specify the Line Spacing and Indents for the **paragraphs** .

5. Modify the Style Properties
Change the Font to Tahoma
Increase the font Size to 16.
Change the color to brown.

There is a **Format** button at the bottom of the screen. Go ahead and see what is there. Return to this screen. Click **OK**.

Review your document: did all of the text formatted as Heading 1 change to Tahoma, 16 pt brown?

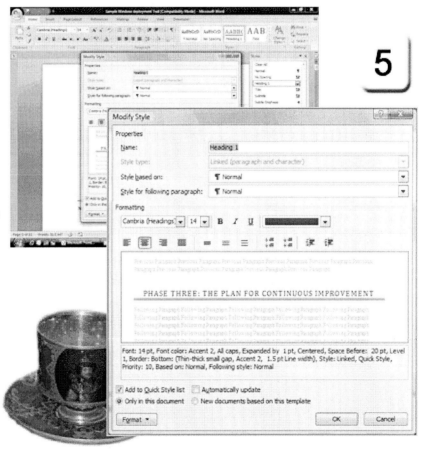

Microsoft Word 2007 Exam 77-601 Topic: **2. Formatting Content**
2.1. Format text and paragraphs
2.1.2. Create and modify styles: Change fonts

Word: Styles Page 1 2 3 4 5 6 7 8 9 10 11 **12** 13 14 15 16 17 18 19 20 21 22 More

Home -> Styles -> Options -> Modify

Quick Style List

Say you **modified an existing Style** to fit your image. How do you want to save your Modified Style?

Look on the bottom for the options. By default, your new Style will be displayed in the **Quick Style list** in this file, only.

You can choose to display your Style in all new documents. It can be added to the Normal template. Each new sheet that opens in Word can use your Style settings instead of Calibri or Times New Roman.

The **Automatically Update** check box can be problematic. Here is a scenario I ran into when I used that option: I'm typing away in Normal and I came across a word that I wanted to emphasize by making Bold. I saw the "busy" sign for several minutes, then all of the Normal text was formatted as Bold. Hmmmm. Well, Undo is my favorite command, sometimes.

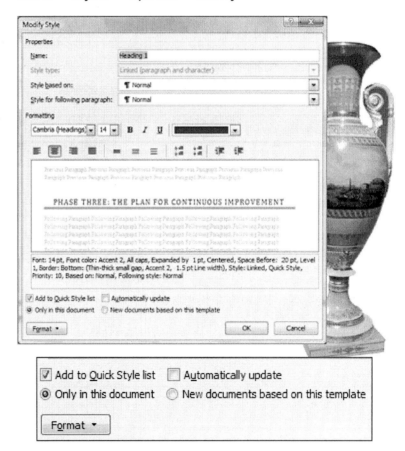

Microsoft Word 2007 Exam 77-601 Topic: 2. Formatting Content
2.1. Format text and paragraphs
2.1.2. Create and modify styles: Create new style based on existing styles

Word: Styles Page 1 2 3 4 5 6 7 8 9 10 11 12 13 14 15 16 17 18 19 20 21 22 More

Home -> Styles -> New Style

Create A New Style
The previous example illustrated how to modify an existing Style. Here are the steps to create a new Style.

Try This: Create Your Own Style
1. Open the Style Pane.
Go to **Home-> Styles**.
Click on the arrow in the bottom right corner to launch the **Styles** Pane.

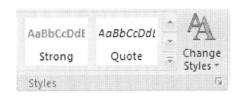

2. Begin a New Style
At the bottom of the Style Pane are three buttons. Click on **New Style**.

Keep going...

Microsoft Word 2007 Exam 77-601 Topic: 2. Formatting Content
2.1. Format text and paragraphs
2.1.2. Create and modify styles: Create new styles

Word: Styles Page 1 2 3 4 5 6 7 8 9 10 11 12 13 **14** 15 16 17 18 19 20 21 22 More

Home -> Styles -> New Style

New Style Formatting

In general, all Styles are based on the **Normal Style**. This goes under the theory that the colors and fonts for the Headings and the Body Text should match: brown suit goes with brown shoes. So begin your new Style by formatting the Normal text.

3. Edit the Style Sheet
Name: My Style.
Type: Paragraph.
Style based on: Normal.

You can choose to add your new Style to the Quick Style gallery.

After you create the Normal style with the font, color and formatting that you want, you would create a **New Style** for Heading 1, 2, and 3 based on My Style.

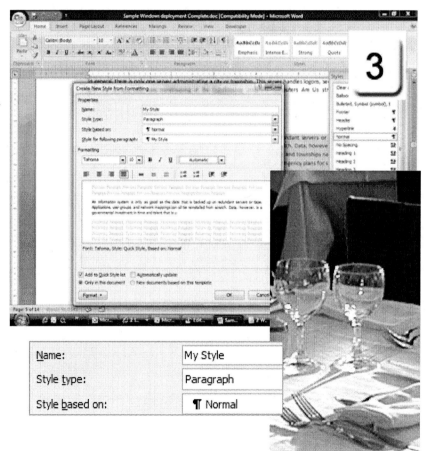

Microsoft Word 2007 Exam 77-601 Topic: 2. Formatting Content
2.1. Format text and paragraphs
2.1.2. Create and modify styles: Create new styles

Word: Styles Page 1 2 3 4 5 6 7 8 9 10 11 12 13 14 15 16 17 18 19 20 21 22 More

Home -> Styles ->Style Inspector

Reveal Style Formatting
Try This: Use the Style Inspector
Select some text that you have formatted with a Style. Open the Style pane. Click on the **Style Inspector**.

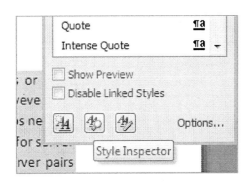

What Do You See? A small Style Inspector will pop up. It should show you the paragraph and text formatting. You can edit or remove the Style if you wish.

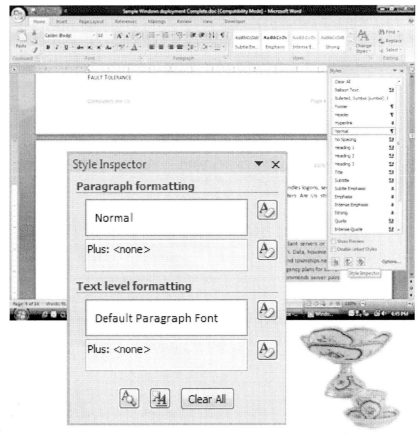

Microsoft Word 2007 Exam 77-601 Topic: 2. Formatting Content
2.1. Format text and paragraphs
2.1.2. Create and modify styles: Reveal style formatting

Word: Styles Page 1 2 3 4 5 6 7 8 9 10 11 12 13 14 15 16 17 18 19 20 21 22 More

Home -> Clipboard ->Format Painter

The Format Painter
The **Format Painter** is a useful little tool that works well. You can use the Format Painter to copy and paste your Styles.

Try It: Use the Format Painter
Select: any **formatted** text.
Go to **Home ->Format Painter**.

What Do You See? The Format Painter looks like a little paint brush. When you select the formatted type and click on Format Painter, your mouse will have a painter brush.

Whatever you "brush" with your mouse will update to the format you picked up from the text you selected at the start.

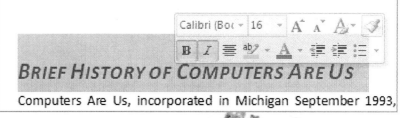

Microsoft Word 2007 Exam 77-601 Topic: 2. Formatting Content
2.1. Format text and paragraphs
2.1.1. Apply styles: Use Format Painter

Word: Styles Page 1 2 3 4 5 6 7 8 9 10 11 12 13 14 15 16 **17** 18 19 20 21 22 More

Home -> Font ->Clear Formatting

Clear the Formatting

Here is a simple way to clear all of the formatting as you work with the Styles and Themes. This tool is also an effective way to remove formatting from information you copied from another webpage or document.

Try This: Clear the Formatting
Go to **Home-> Font.**
Select: **Clear Formatting.**

What Do You See? When you remove the formatting the text will return to the default. In Word 2007 the **Font** is Calibri 11 pt.

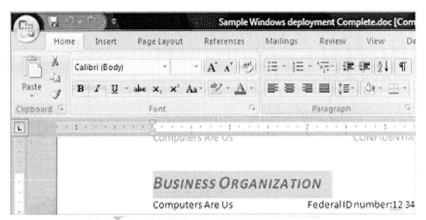

Word: Styles Page 1 2 3 4 5 6 7 8 9 10 11 12 13 14 15 16 17 18 19 20 21 22 More

Insert -> Page Break

Take a Break

One of the benefits of working with Styles is how quickly you can create a Table of Contents. There are two steps you need to complete before you can work with the T of C: page breaks and page numbers.

The first four lines of this document should go on the cover page. Raise your hand if you force the rest of the text to a new page by adding a bunch of blank lines? This is not your best solution. If you add more text to the first page, the rest of paragraphs will shift down, too.

6. Insert a Page Break
Go to page 1
Place your cursor after the date
Go to **Insert ->Page Break**

Word: Styles Page 1 2 3 4 5 6 7 8 9 10 11 12 13 14 15 16 17 18 **19** 20 21 22 More

Insert -> Break

Insert A Break

If you use the **Show/Hide** button, you can see the code for the Page Break. If you want to remove the break, select the Page Break code and delete it.

Additional page breaks:
Please insert page breaks before the following headlines:
Project Summary
Phase 1
Phase 2
Phase 3

Memo to Self: You can turn off Show/Hide after you remove the Page Break if you like.

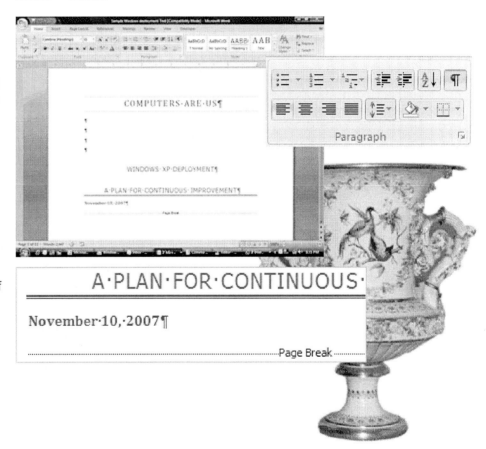

Microsoft Word 2007 Exam 77-601 Topic: 2. Formatting Content
2.3. Control pagination
2.3.1. Insert and delete page breaks

Word: Styles Page 1 2 3 4 5 6 7 8 9 10 11 12 13 14 15 16 17 18 19 20 21 22 More

View -> Document Map

View the Document Map

Consistent headlines are one benefit of using Styles. Another hidden treasure in Microsoft Word is the **Document Map**.

Try it: View the Document Map
Go to the **View** tab
Select **Document Map**

You should see an outline along the left side of your document. This outline includes each of the headlines you formatted with a Style.

When you click on a headline in the Document Map, you are instantly taken to that place. This can be very useful way to navigate around a 20 page proposal!

Click on **Document Map** and switch to **Thumbnails**. What do you think?

Close the Document Map by clicking on the small **X**.

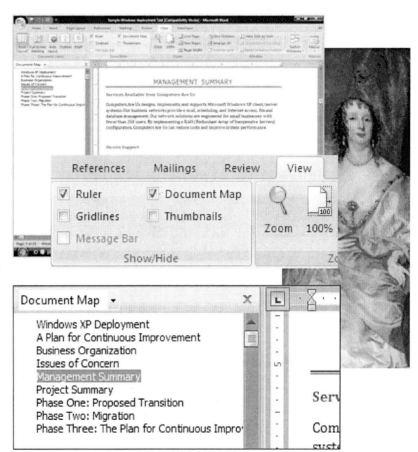

Word: Styles Page 1 2 3 4 5 6 7 8 9 10 11 12 13 14 15 16 17 18 19 20 **21** 22 More

Insert -> Header and Footer

Headers and Footers

This is a good time to add the page numbers so that you can create a Table of Contents. Page Numbers can be placed in the Header or the Footer.

7. Insert a Footer

Go to page 3, Business Summary
Go to **Insert ->Footer.**

Select one of the **Footer** templates that includes page numbers. The templates set up the Tab Stops. You can use the **Page Number** button if you wish.

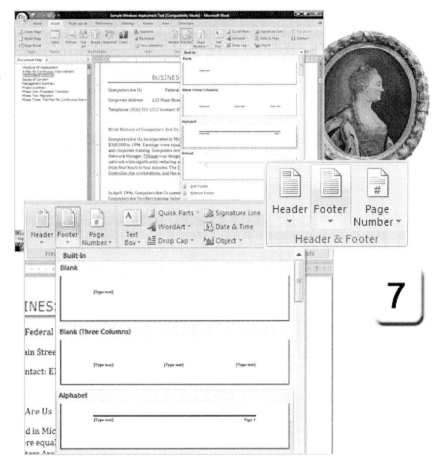

Microsoft Word 2007 Exam 77-601 Topic: 1. Creating and Customizing Documents
1.2. Lay out documents
1.2.1. Format pages: Page numbers

Word: Styles Page 1 2 3 4 5 6 7 8 9 10 11 12 13 14 15 16 17 18 19 20 21 22 More

Header and Footer Tools -> Design

Headers and Footers
A special text box will open on the top and bottom of your document. The rest of your document text will appear dim, and cannot be edited while you work with the Headers and Footers.

Try it: Edit the Footer
Go to the **Footer**
Click on **Type Text**
Type: Workstation Upgrade

You can also insert **Page Numbers** in the Footer.

Microsoft Word 2007 Exam 77-601 Topic: 1. Creating and Customizing Documents
1.2.2. Create and modify headers and footers (Not using Quick Parts)

Word: Styles (page 1) 22 23 24 25 26 27 28 29 30 31 32 33 34 35 36 37 38 39 40 41 42 43 44 45 46 47 48

Header and Footer Tools -> Design ->Date and Time

Modify the Date and Time

Many documents also need a Date/Time Stamp. It can be added to the Footer as well. Did you know that you make the Date/Time automatically update?.

Try This: Update the Date and Time
Go to **Header and Footer Tools.**
Select the **Design** tools.
Click on **Date and Time.**

What Do You See? A new window will offer several short, medium and long date formats. Look in the bottom right corner for the option to **Update Automatically.**

How Does It Work? When you click on the Date/Time you should see the **Update** option. Pretty cool.

Word: Styles (page 1) 22 23 24 25 26 27 28 29 30 31 32 33 34 35 36 37 38 39 40 41 42 43 44 45 46 47 48

Add a Cover Page

Professional documents, such as a Request for Proposal or Quarterly Report generally have a **Cover Page**.

Try This: Insert a Cover Page
Place your cursor at the top of the first page in your sample document.
Go to **Insert-> Cover Page.**

Insert -> Cover Page

What Do You See? Word 2007 has a gallery of Cover Page templates. These templates have interesting Quick Parts or you can use your custom Quick Parts if you wish.

Microsoft Word 2007 Exam 77-601 Topic: 1. Creating and Customizing Documents
1.1.6. Insert blank pages or cover pages

Word: Styles (page 1) 22 23 24 25 26 27 28 29 30 31 32 33 34 35 36 37 38 39 40 41 42 43 44 45 46 47 48

More Headers and Footers
Did you know that you can have different Headers and Footers for each part of your document? Start with the Cover Page. Usually the Cover does not have any page numbers.

Try it: Create a Different First Page
Start on Page 3.
Go to the **Header and Footer Tools.**
Click on the **Design** tab.
Check **Different First Page.**

Try This, Too: Confirm the changes
Go to **View** and Zoom to Two Pages.

Is the **First Page Footer** different than the Footer on page 2?

Header and Footer Tools -> Design-> Options

View -> Zoom-> Two Pages

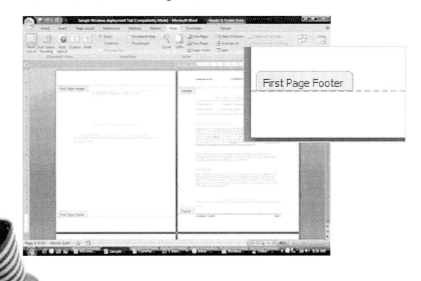

Word: Styles (page 1) 22 23 24 25 26 27 28 29 30 31 32 33 34 35 36 37 38 39 40 41 42 43 44 45 46 47 48

Page Layout -> Breaks ->Section Breaks

Different Sections

Say you had three **Sections** in your document and you wanted each section to have a different Header or Footer. To do this, you need to insert a **Break** and choose a **Continuous Section Break**.

There are four kinds of Section Breaks: Next Page, Continuous, Even and Odd.

Try it: Insert A Section Break
Begin on page 3.
Place your cursor before the headline: Issues of Concern.

Go to **Page Layout**.
Find the **Page Setup** Group.
Select **Breaks**.
Choose **Section Breaks ->Continuous**.

Add a Continuous Section Break to:
Management Summary
Phase 1
Phase 2
Phase 3

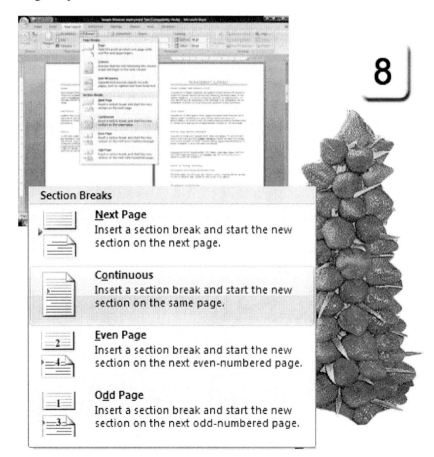

Word: Styles (page 1) 22 23 24 25 26 27 28 29 30 31 32 33 34 35 36 37 38 39 40 41 42 43 44 45 46 47 48

Header and Footer Tools -> Design-> Navigation

Different Footers

Working with the sections requires some thought. The Computer Mama cusses the computer if it requires too much thinking...

The key to making this option work is the **Link to Previous** command. When that link to turned off, you can make each Section have unique Headers and Footers.

Try it: Remove Link to Previous
Go to page 3, Issues of Concern.
Double click the **Header** to open it.
Type: Computer Are Us.
Tab to the center of the Header.
Type: Issues of Concern.

Go to **Header and Footer Tools -> Design**
Go to the **Next Section**, Phase 1.
Do NOT Link to Previous.
Type: Phase 1.

Try it with the Phase 2 and Phase 3 sections, too.

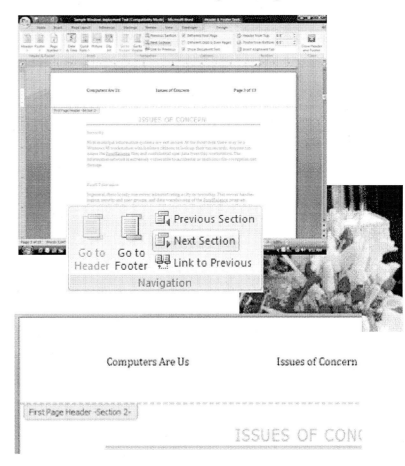

Word: Styles (page 1) 22 23 24 25 26 27 **28** 29 30 31 32 33 34 35 36 37 38 39 40 41 42 43 44 45 46 47 48

Home -> Paragraph -> Show/Hide

Delete the Section Breaks
Try This: Remove the Breaks
Go to **Home -> Paragraph**.
Click on the **Show/Hide** button.

What Do You See? Each **Page Break** and **Section Break** is identified. You can place your cursor on the Break and delete it if you wish.

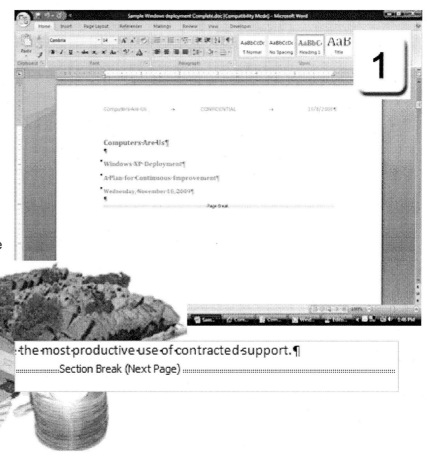

Word: Styles (page 1) 22 23 24 25 26 27 28 29 30 31 32 33 34 35 36 37 38 39 40 41 42 43 44 45 46 47 48

References -> Table of Contents

Table of Contents
Now that you have the Styles in place, it is really simple to build a **Table of Contents**.

9. Create A Table of Contents
First, make a blank page between the Cover Sheet and Page 2. Place your cursor at the top of Page 2 and go to Page Layout. Insert a Page Break.
Go to the new, blank Page 2,
Go to **References.**
Select **Table of Contents.**

Memo to Self: You can add a Table of Contents, Figures or Authority with Quick Parts, too.
Go to **Insert->Quick Parts**.
Select: **Field**.

Look through the Field list and you will recognize many of the tools in this lesson.

Microsoft Word 2007 Exam 77-601 Topic: 1. Creating and Customizing Documents
1.3. Make documents and content easier to find
1.3.1. Create, modify and update tables of contents

Word: Styles (page 1) 22 23 24 25 26 27 28 29 <u>30</u> 31 32 33 34 35 36 37 38 39 40 41 42 43 44 45 46 47 48

References -> Table of Contents -> Insert Table of Contents

Table of Content Options

The entries in the Table of Contents are generated from the text formatted as Heading 1. You can change the Table if you wish.

Try it: Modify the T of C
Go to **References**.
Select **Table of Contents**.
Go to the bottom of the menu and choose **Insert Table of Contents**.

You should see the options for page numbers, alignment, Tab leader, and how many levels you want.

You will be prompted to replace the first table that you made earlier. Click OK to see the revised T of C.

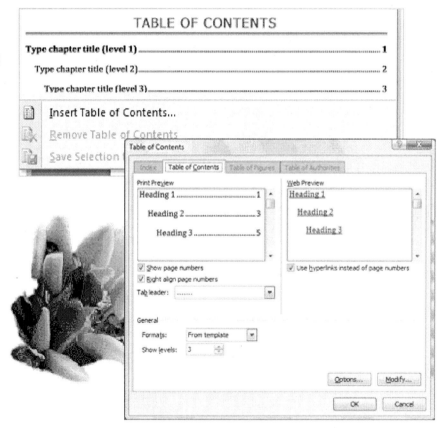

Word: Styles (page 1) 22 23 24 25 26 27 28 29 30 31 32 33 34 35 36 37 38 39 40 41 42 43 44 45 46 47 48

References -> Table of Contents -> Update Table

Editing the T of C

The Table of Contents is really a set of hyperlinks to your Headings. Run your mouse over the T of C and watch the cursor. Yes, you can click on any topic and jump to the right page.

The T of C can be edited like text. You can highlight lines of type and delete references if you want to. Say you added more information to one of the pages and all of your text moved to the next page. How do you update the Table of Contents?

Try it: Update the Table of Contents
Select the Table of Contents.
Go to the **References** tab.
Select **Update** from the options.

You can revise just the page numbers, or you can modify the entire table if you added more headings.

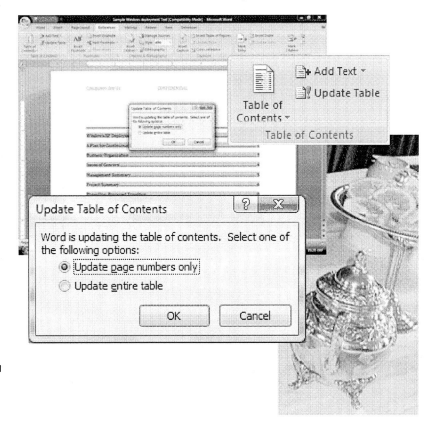

Word: Styles (page 1) 22 23 24 25 26 27 28 29 30 31 <u>32</u> 33 34 35 36 37 38 39 40 <u>41</u> 42 43 44 <u>45</u> 46 47 48

References -> Table of Contents -> Add Text

Add Text to the T of C

The **Reference** Ribbon has an effective tool for formatting text and automatically adding it to the T of C.

Try it: Add Text to the T of C
Select a few words of text in your sample document.
Go to the **References** Ribbon.
Select **Add Text** from the options.

What Do You See? You can select Level 1, 2 or 3. The Levels in the Table of Contents are formatted with your headline Styles.

You do not have to include this text in the Table of Contents..

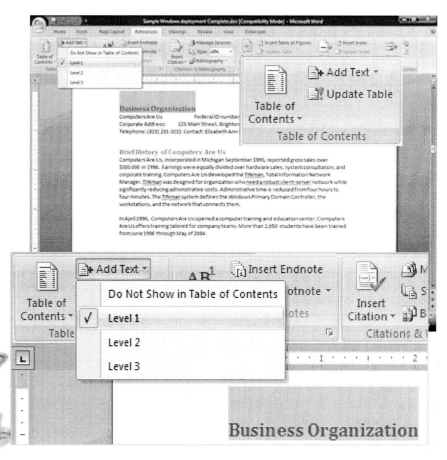

Microsoft Word 2007 Exam 77-601 Topic: 1. Creating and Customizing Documents
1.3. Make documents and content easier to find
1.3.1. Create, modify and update tables of contents: Update tables of contents with selected text

Word: Styles (page 1) 22 23 24 25 26 27 28 29 30 31 32 **33** 34 35 36 37 38 39 40 41 42 43 44 45 46 47 48

Mark Text for an Index

Creating an Index is similar to creating a Table of Contents.

Try This: Mark Text for the Index
Select the text your would like to include in the Index.
Go to **Reference -> Index**.
Select: **Mark Entry**.

Reference -> Index ->Mark Entry

What Do You See? The text you selected will be the **Main entry**. You can add a **Subentry** or **Cross-reference** if you wish. When you **Mark** an entry, the Show/Hide will reveal the Index code: {XE "Sample Text"}.

Repeat these steps for each topic that you want to include in the Index.

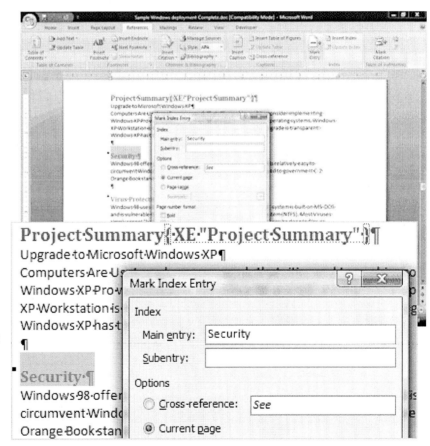

Microsoft Word 2007 Exam 77-601 Topic: 1. Creating and Customizing Documents
1.3. Make documents and content easier to find
1.3.2. Create, modify and update indexes: Mark an entry for indexing

Word: Styles (page 1) 22 23 24 25 26 27 28 29 30 31 32 33 34 35 36 37 38 39 40 41 42 43 44 45 46 47 48

Reference -> Index ->Insert Index

Create the Index
After you mark your text for entry, you can create the Index.

Try This: Insert the Index
An Index is usually at the end of document. Please place your cursor at the end and insert a blank page.
Go to **Reference -> Index**.
Select: **Insert Entry**.

What Do You See? By default, the Index is indented and formatted for 2 columns. The new Index will show the marked entries as an alphabetical list that includes the page numbers.

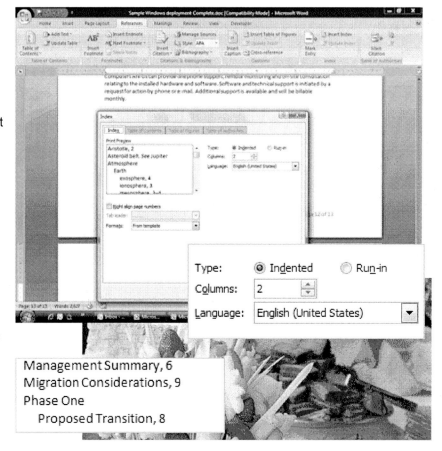

Microsoft Word 2007 Exam 77-601 Topic: 1. Creating and Customizing Documents
1.3. Make documents and content easier to find
1.3.2. Create, modify and update indexes

Word: Styles (page 1) 22 23 24 25 26 27 28 29 30 31 32 33 34 35 36 37 38 39 40 41 42 43 44 45 46 47 48

Reference -> Citations and Bibliography -> Insert Citation

Insert a Citation

A **Bibliography** is a list of the sources and references that you used in your report or paper. The Bibliography is compiled by marking Citations throughout your document.

Try This: Insert a Citation
Place your cursor at the end of a sentence in your sample document.
Go to the **Reference Ribbon**.
Click on: **Insert Citation**.
Fill in the blanks.

What Do You See? The Citation in this example reads (Spolsky, 2000).

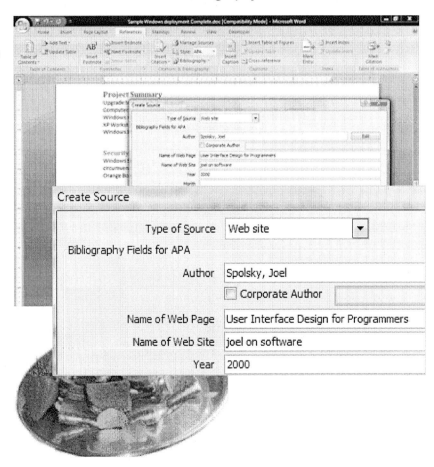

Microsoft Word 2007 Exam 77-601 Topic: 4. Organizing Content
4.4. Insert and format references and captions
4.4.2. Insert citations and captions

Word: Styles (page 1) 22 23 24 25 26 27 28 29 30 31 32 33 34 35 36 37 38 39 40 41 42 43 44 45 46 47 48

Reference -> Citations and Bibliography

Working with Citations

What Do You See? The Citation in this example reads (TechNet, 2008).

Try This: Edit the Citation
Click on the Citation. The options are:
Edit the citation
Edit the source
Convert to text

If you edit your citation you can use this control to **Update the Bibliography** as well as the Citation reference in your Master List.

Memo to Self: Citations have new functionality in Word 2007. These functions are not available in documents created in previous versions of Word. You can copy and paste your work into a new Word 2007 document to enable these options.

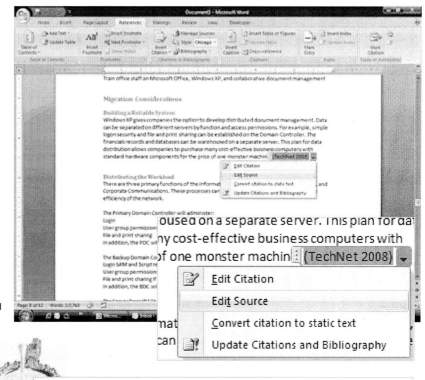

Word: Styles (page 1) 22 23 24 25 26 27 28 29 30 31 32 33 34 35 36 37 38 39 40 41 42 43 44 45 46 47 48

Reference -> Citations and Bibliography ->Style

Citation Styles

There are several reference styles in Microsoft Word 2007. The list includes the **Chicago Manual of Style**, a veteran style reference that has been published by the University of Chicago Press since 1906. The **MLA** is another reference style that is published by the Modern Language Association.

The **APA style** is published by the America Psychology Association and is used for social science papers and dissertations. The APA style uses <u>Harvard referencing</u>: the author-date system of citations and parenthesis.

Try This Too: Create a Placeholder
You can create a Citation placeholder and fill in the source later if you wish.

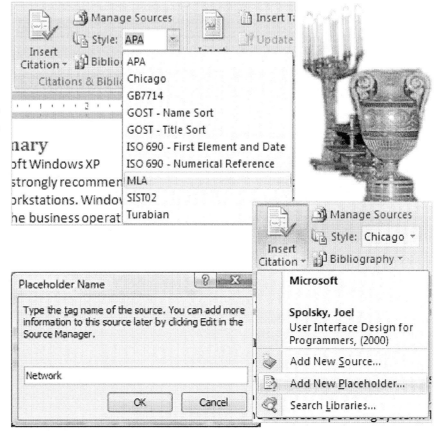

Microsoft Word 2007 Exam 77-601 Topic: 4. Organizing Content
4.4. Insert and format references and captions
4.4.4. Select reference styles: Choose MLA, APA, or Chicago Manual of Style

Word: Styles (page 1) 22 23 24 25 26 27 28 29 30 31 32 33 34 35 36 37 <u>38</u> 39 40 41 42 43 44 45 46 47 48

Reference -> Citations and Bibliography ->Manage Sources

Managing Sources

Microsoft Word 2007 has a convenient method for organizing your sources. You can use **Manage Sources** to find a reference by author, title, year or tag. You can also use this tool to transfer your sources to another document.

Try This: Manage Your Sources
Go to **References Ribbon**.
Click on: **Manage Sources.**

What Do You See? The Master List displays all of the sources and placeholders available. You can Copy, Delete, Edit and Create a New source.

Memo to Self: You can edit your **Placeholders** in this **Manage Sources** window.

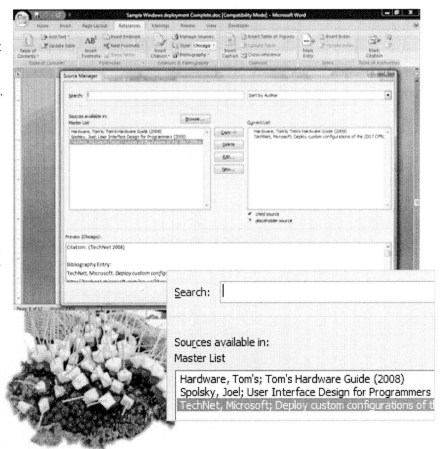

Microsoft Word 2007 Exam 77-601 Topic: 4. Organizing Content
4.4. Insert and format references and captions
4.4.1. Create and modify sources

Word: Styles (page 1) 22 23 24 25 26 27 28 29 30 31 32 33 34 35 36 37 38 **39** 40 41 42 43 44 45 46 47 48

Insert a Bibliography
After you insert the **Citations**, you can create a **Bibliography**. Here are the steps.

Try This: Insert a Bibliography
Please place your cursor on the last page of your sample document.
Go to **Reference -> Bibliography**.

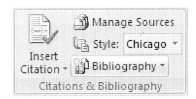

What Do You See? When you click on the Bibliography, you will see the options to **Format** or **Update Citations and Bibliography**.

Way cool new technology.

Reference -> Bibliography

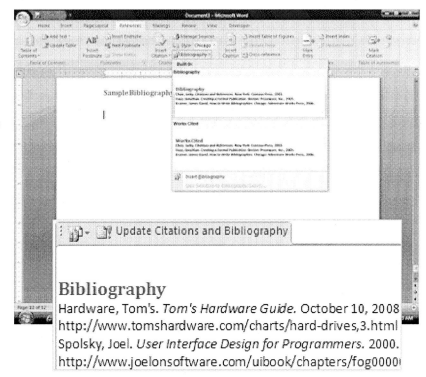

Word: Styles (page 1) 22 23 24 25 26 27 28 29 30 31 32 33 34 35 36 37 38 39 40 41 42 43 44 45 46 47 48

Reference -> Insert Caption

Add a Caption to a Picture
Each task begins by marking, or tagging, the information you wish to include in your list. Creating a **Table of Figures** is similar to selecting text for an Index.

Before You Begin: Add Some Pictures
Select a place in your sample document.
Go to **Insert-> Picture.**
Select and resize two or three pictures from your sample pictures or ClipArt.

Try This: Add a Caption to a Picture
Select a picture.
Go to **Reference -> Insert Caption**.
Fill in the blanks.

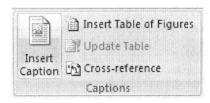

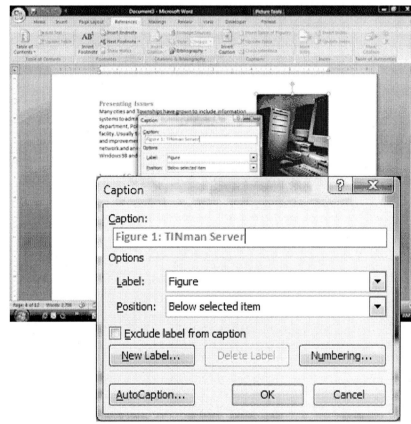

Keep going...

Word: Styles (page 1) 22 23 24 25 26 27 28 29 30 31 32 33 34 35 36 37 38 39 40 41 42 43 44 45 46 47 48

Reference -> Insert Table of Figures

Insert a Table of Figures
Say you selected several pictures and added a caption to each one. Now, you can create a **Table of Figures**.

Try This: Insert a Table of Figures
Place your cursor at the end of your sample document.
Go to the **Reference Ribbon.**
Go to **Insert Table of Figures**.

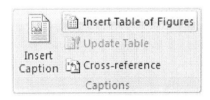

What Do You See? You can choose to show the page numbers, edit the alignment, and select a Tab leader.

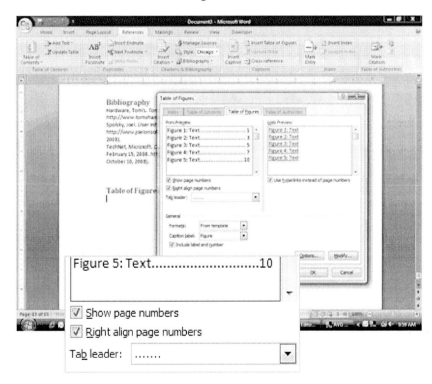

Microsoft Word 2007 Exam 77-601 Topic: 4. Organizing Content
4.4. Insert and format references and captions
4.4.5. Create, modify and update tables of figures and tables of authorities

Word: Styles (page 1) 22 23 24 25 26 27 28 29 30 31 32 33 34 35 36 37 38 39 40 41 42 43 44 45 46 47 48

Home -> Styles ->Intense Quote

Format Quoted Material
Here is a bit of formatting that's fun to use. You can quote me on that...
OK, bad pun.

Before you Begin: Enter a Quote
Place your cursor in the body text of your sample document. Type:
"Controlling your environment makes you happy. "

Try This: Format Quoted Material
Select the text you just typed in.
Select the Go to **Home -> Styles**.
Select: **Intense Quote**.

Try This: Format a Paragraph
If you select a paragraph, the Quote Style will indent the tab stops equally from the left and right margins.

Now, you're just showing off. ;-)

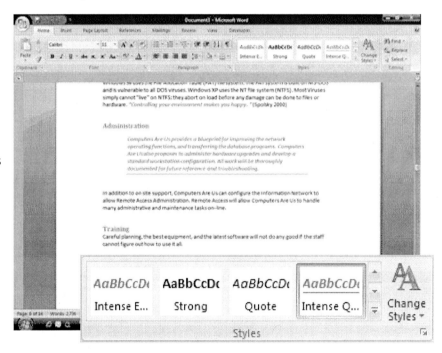

Microsoft Word 2007 Exam 77-601 Topic: 2. Formatting Content
2.1. Format text and paragraphs
2.1.4. Format paragraphs: Format quoted material

Word: Styles (page 1) 22 23 24 25 26 27 28 29 30 31 32 33 34 35 36 37 38 39 40 41 42 **43** 44 45 46 47 48

Insert -> Drop Cap

Show Me: Drop Caps
A **Drop Cap** is a distinctive method to format Text as Art. When you insert a Drop Cap you are creating a little text box that can be formatted inside and out.

You can select the Font for your Drop Cap letter You can also format the Text Wrap around the Text Box.

Try This: Insert a Drop Cap
Select the First letter in any paragraph in your sample document.
Go to **Insert-> Drop Cap.**
Select **Dropped**.

Please keep going...

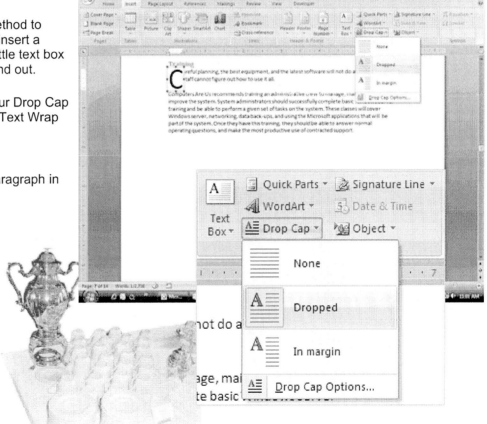

Word: Styles (page 1) 22 23 24 25 26 27 28 29 30 31 32 33 34 35 36 37 38 39 40 41 42 43 44 45 46 47 48

Insert -> Drop Cap ->Drop Cap Options

Drop Cap Options

What Do You See? How do you want the text to wrap around your Drop Cap? You can format the **Position**.

If you select a **Dropped Position** you can specify how many lines you wish. The more lines you choose, the bigger your Drop Cap letter.

You can also format the **Font**.

Memo To Self: In hand made books in olden days, the Drop Cap was often a very fancy type face. What do you think? Is there a Font that works with your corporate image?

Word: Styles (page 1) 22 23 24 25 26 27 28 29 30 31 32 33 34 35 36 37 38 39 40 41 42 43 44 **45** 46 47 48

Home -> Editing ->Find and Replace

Find and Replace Text
Headlines, Styles and Reference Tables make it easier to navigate a document. The formatting acts like landmarks, confirming that you are on the right page. Microsoft Word also has several editing tools on the Home Ribbon that can assist with navigation or manipulating text.

Try This: Find Specific Text
Go to **Home-> Editing.**
Click on the **Find** button.

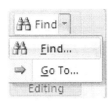

What Do You See? The **Find and Replace** window will open. Go to the **Find** tab and type in the search box Word will highlight the text you are seeking.

Keep going....

Microsoft Word 2007 Exam 77-601 Topic: 2. Formatting Content
2.2. Manipulate text
2.2.2. Find and replace text: Search for and highlight specific text

Word: Styles (page 1) 22 23 24 25 26 27 28 29 30 31 32 33 34 35 36 37 38 39 40 41 42 43 44 45 46 47 48

Home -> Editing ->Find and Replace

Find and Replace Options

Find and Replace can take this process one step further. After you locate specific text, you can replace it. This is an excellent way to recycle a report and update the content.

Try This: Find and Replace
Go to **Home-> Editing.**
Click on the **Replace** button.

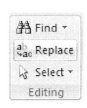

What Do You See? If you use **Replace**, you can review each highlighted selection one at a time. You can also choose **Replace All,** and have Microsoft Word automatically update all of the text that matches your criteria, if you wish.

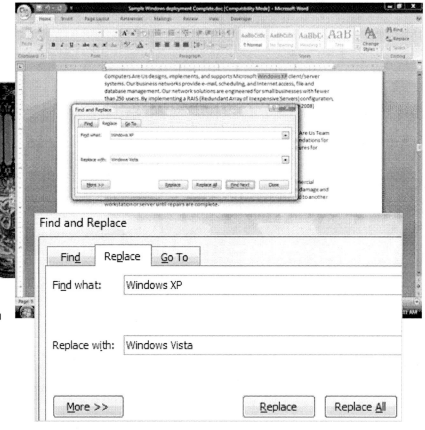

Microsoft Word 2007 Exam 77-601 Topic: 2. Formatting Content
2.2. Manipulate text
2.2.2. Find and replace text: Replace all

Word: Styles (page 1) 22 23 24 25 26 27 28 29 30 31 32 33 34 35 36 37 38 39 40 41 42 43 44 45 46 47 48

Home -> Editing ->Go to

Find and Go To Options

Another way to locate information in your document is to use the **Go To** command. With **Go To**, you can specify what you are looking for. It is a sophisticated Search tool for navigating your work.

You can go to any page, section, line, bookmark, comment or reference.

Try This: Review the Go To Options
Go to **Home-> Editing**.
Click on **Find -> Go To**.

Try This, Too: Select All
The Select option lets you:
Select All, Select Objects, or click on a headline and Select Text with Similar Formatting. Very good...

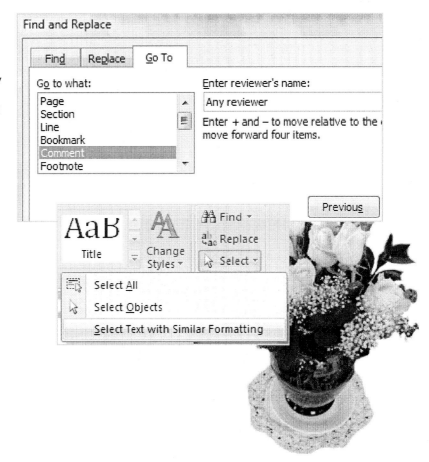

Microsoft Word 2007 Exam 77-601 Topic: 5. Reviewing Documents
5.1. Navigate documents
5.1.1. Move a document quickly using the Find and Go To commands

Word: Styles (page 1) 22 23 24 25 26 27 28 29 30 31 32 33 34 35 36 37 38 39 40 41 42 43 44 45 46 47 48

DONE

Doing It With Style

Each of the topics in this lesson focused on tools with functionality: Styles, Quick Parts, and References are all programmable. They can be updated with one click if you wish.

This is how computers are supposed to behave: not friendly, just obedient.

Save your work. Done and Done. You get the cookie. <grin>

1. Which of the following can be inserted in the Insert-> Pages group? (Select all correct answers.)
a. Cover Page
b. Blank Page
c. Page Break
   Tip: Advanced Guide to Word, p64

2. To give each section in a document a different header, the Link to Previous option has to be turned...
a. On
b. Off
   Tip: Advanced Guide to Word, p67

3. To create an index, first you mark the text that you want included in the index.
a. TRUE
b. FALSE
   Tip: Advanced Guide to Word, p73

4. Word2007 includes a tool for creating citations where the user fills in the boxes in the Create Source dialogue box.
a. TRUE
b. FALSE
   Tip: Advanced Guide to Word, p75

5. Where does the Bibliography function get the list of sources?
a. The user must enter them by hand
b. From the Citations entered in the text
c. From the Internet
   Tip: Advanced Guide to Word, p75

6. Which of the following Citation Styles are available in Word2007? (Select all correct answers.)
a. APA
b. Chicago Manual of Style
c. MLA
d. Microsoft
   Tip: Advanced Guide to Word, p77

7. Which of the following are true about Captions for figures in Word?
a. They are used in a Table of Figures
b. They can be displayed below the figure
c. The command is on the Reference Ribbon, in the Caption group
   Tip: Advanced Guide to Word, p80, 81

Page 1 2 3 4 5 6 7 8 9 10 11 12 13 14 15 16 17 18 19 20 21 22 23 More

The Good, The Bad, and the Really Ugly

Who Done It?

Click Here to Get Started
Sample Files

Advanced Word

Lesson Objectives: Learn how to create an interactive form. This lesson will demonstrate how to organize content with tables, control document access, customize themes, track changes and compare documents. In this lesson you will:

Use the Insert Ribbon to add a Table page 3

Identify the steps needed to show the Developer Toolbar page 6

Use the Developer Toolbar to add Controls page 7

Learn why you need to Protect the Document page 11

Practice using Themes to format the form page 14

Identify and use the Document Inspector page 24

Use the Review Ribbon to Track Changes page 28

Use the Review Ribbon to find the Research options page 42

Learn how to customize Word 2007 page 45

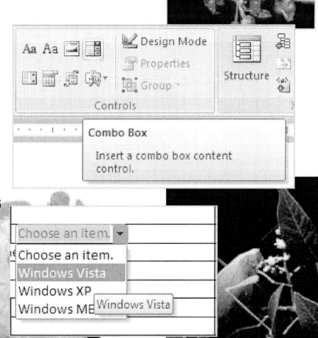

Word: Who Done It? Page 1 2 3 4 5 6 7 8 9 10 11 12 13 14 15 16 17 18 19 20 21 22 23 More

Creating Forms in Word

Every office—big or small—needs to generate forms. We use a form to gather data and to insure that the information is complete. Today we're going to use Microsoft Word to create a form that we can answer on-line. Our form is going to be a status report for when a computer or application fails. So, let's **Start** the Program **Microsoft Word**.

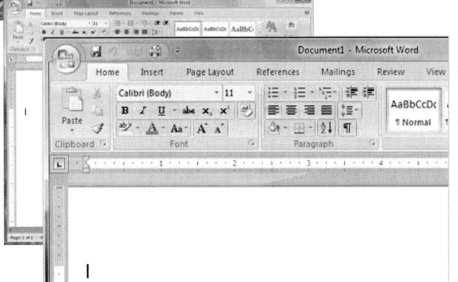

What do you see, from the top of the screen? Is there a **Title** Bar that says Microsoft Word? Yes.

Is there a **Home** Ribbon with the Clipboard, Font and Paragraph Groups? Yes.

If your screen looks similar to the example on this page, then you are ready to get started.

Word: Who Done It? Page 1 2 3 4 5 6 7 8 9 10 11 12 13 14 15 16 17 18 19 20 21 22 23 More

Insert ->Table

Use a Table for Form Design

You can use the columns and rows in a table to make a professional form and simplify the form design.

1. Insert a Table

Go to the **Insert Ribbon** and click on **Table**. Now, highlight a 3x8 table.

You should see a grid in your document. The upper left hand corner square is called cell A1, same as the first cell in an Excel spreadsheet.

Word: Who Done It? Page 1 2 3 4 5 6 7 8 9 10 11 12 13 14 15 16 17 18 19 20 21 22 23 More

Insert ->Table

Enter the Labels

2. Type the Labels
Click in the first Cell, A1, and type: Name
Tab to the third Cell, C1, and type: Location

Row 2: When you tab again your cursor will go to the Cell A2 in the second row.
Click in Cell A2 and type: Computer Name
Tab to the third Cell, C2 and type: Date

Row 3: Go to Row 3
Click in Cell A4 and type: Operating System

Row 4: Go to Row 5
Click in Cell A5 and type: Applications Open at the Time of the Incident

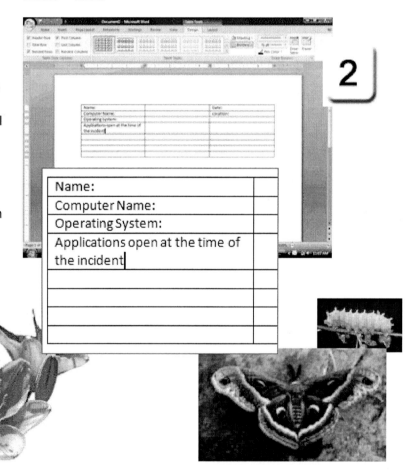

Microsoft Word 2007 Exam 77-601 Topic: 4. Organizing Content
4.2. Use tables and lists to organize content

Word: Who Done It? Page 1 2 3 4 5 6 7 8 9 10 11 12 13 14 15 16 17 18 19 20 21 22 23 More

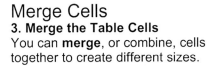

Table Tools -> Layout -> Merge

Merge Cells

3. Merge the Table Cells
You can **merge**, or combine, cells together to create different sizes.

Select Cell A1 and Cell B1, the one with the label "Name" and the next one.
Go to the **Table Tools**
Click on the **Layout** tab
Select **Merge**

Select Cells A3 and B3.
Go to Table T**ools->Layout -> Merge**

Select Cells A4 and B4.
Go to **Table Tools->Layout -> Merge**
Do you still see the label: Applications Open at the Time of the Incident?

Select Cells A5 and B5.
Go to **Table Tools->Layout -> Merge**
In Cell A6 type: Severity of the Problem

Select Cells A6 and B6.
Go to **Table Tools->Layout -> Merge**
In Cell A7 type: Presenting Issue

Microsoft Word 2007 Exam 77-601 Topic: 4. Organizing Content
4.3. Modify tables
4.3.3. Merge and split table cells

Word: Who Done It? Page 1 2 3 4 5 6 7 8 9 10 11 12 13 14 15 16 17 18 19 20 21 22 23 More

Office -> Popular -> Show Developers tab

Developer Toolbar

After you save your document, try to enter your name. Did you notice that it was not very easy to get from the Name to the Location field?

An on-line form needs to be more "user friendly." People should start with their cursor in the Name box. When they hit the Tab key, they should move to the next box and fill in the blanks.

4. Open the Developer's Toolbar
To set up this experiment, we need to get another tool bar.

Go to **Office** and select **Options**

Go to the **Popular** tab
Check **Show Developers tab**

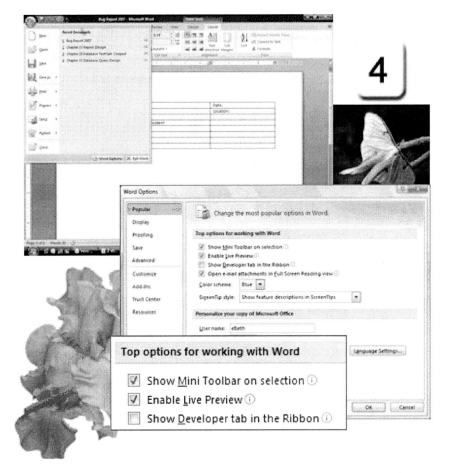

Microsoft Word 2007 Exam 77-601 Topic: 1. Creating and Customizing Documents
1.4. Personalize Office Word 2007
1.4.1. Customize Word options

Word: Who Done It? Page 1 2 3 4 5 6 7 8 9 10 11 12 13 14 15 16 17 18 19 20 21 22 23 More

Developer -> Controls -> Rich Text Control

Rich Text Control
5. Create a Text Control
Place your cursor to the right of the "Name" label.

Go to the **Developer** tab.
Find the **Controls** group.
Click the **Text Form Field (Aa)**.
There are two **Text Controls**. One is for Plain text, the other is for Rich text-big, bold, blue.

You should see a new **Control** that instructs the user: Click here...

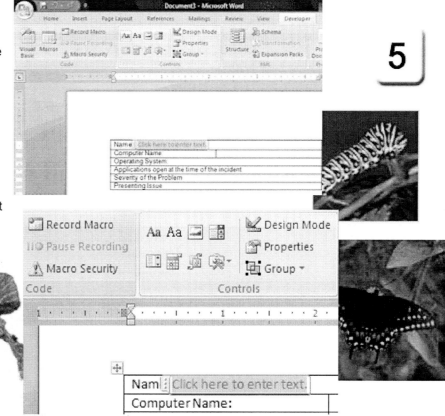

Word: Who Done It? Page 1 2 3 4 5 6 7 8 9 10 11 12 13 14 15 16 17 18 19 20 21 22 23 More

Developer -> Controls -> Date Picker

Date Picker Control
6. Add the Date Form Field
Click to the right of the "Date" label. Select the **Date Picker** Control.

Try it: Tell me that's not the coolest thing you've done in a while.

Word: Who Done It? Page 1 2 3 4 5 6 7 8 **9** 10 11 12 13 14 15 16 17 18 19 20 21 22 23 More

Combo Box Control
Selecting answers from a list means the data will be consistent and it will take less time to fill out the form. There are only so many Operating Systems, so that would be a good place for a **Combo Box Control**.

7. Add a Combo Box Control
Go to the **Developer** tab
Select **Combo Box** from the Control group

You should see a new Control that prompts you to Choose an item.

Change the Combo Box Properties
If you try to use the Combo Box right now, there is nothing on the list to choose. You can add items by editing the Properties.

Try it: Select the Combo Box
Go to the **Developer** tab
Select **Properties**

Developer -> Controls -> Combo Box

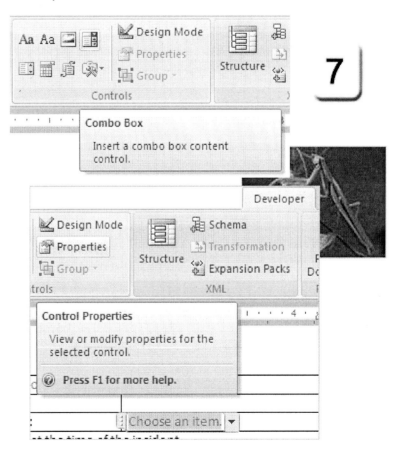

Word: Who Done It? Page 1 2 3 4 5 6 7 8 9 <u>10</u> 11 12 13 14 15 16 17 18 19 20 21 22 23 More

Developer -> Controls -> Properties

Combo Properties
8. Edit the Combo Box Properties
Enter a **Title** for the Combo Box.
Do NOT check Contents cannot be edited.
Add to the Drop-Down List
Type a **Display Name** and **Value**

Add the follow sample items to the list.

Display Name: Windows Vista
Value: Vista

Display Name: Windows XP
Value: XP

Display Name: Windows ME
Value: ME2

The Display Name is what you will see in the list when you click on the Combo Box. The Value is the information stored when you choose an item from the list.

You can shuffle the items in the list by selecting it and using the Move Up or Move Down options.

Word: Who Done It? Page 1 2 3 4 5 6 7 8 9 10 **11** 12 13 14 15 16 17 18 19 20 21 22 23 More

Developer -> Protect Document

Protect the Document
A form will not work properly until you **Protect the Document**. This places the form in User mode, and turns off the editing. Protecting the document also activates the form Controls.

9. Protect the Document
Go to the **Developer** tab
Select **Protect Document**

You will see a new Task bar on the right side of your document.

Check **Editing Restrictions**
Select **Filling in Forms**

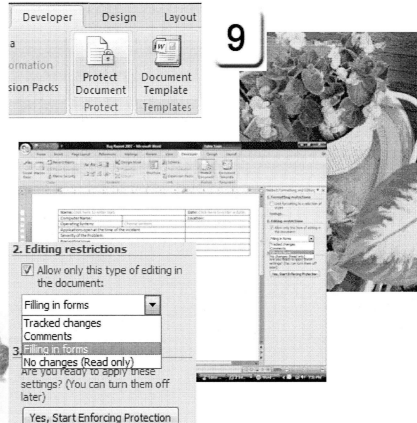

Microsoft Word 2007 Exam 77-601 Topic: 6. Sharing and Securing Content
6.2. Control document access
6.2.1. Restrict permissions to documents

Word: Who Done It? Page 1 2 3 4 5 6 7 8 9 10 11 **12** 13 14 15 16 17 18 19 20 21 22 23 More

Developer -> Protect Document

Test the Form
Well, does it work? Microsoft Word Forms have to be **Protected** before they work. When the document is locked, a user can only fill in the blanks. They can't edit the labels or change the Controls. Try it.

You can add a password when you start the enforcement. This is not real security, but it will stop most curious users from altering your form.

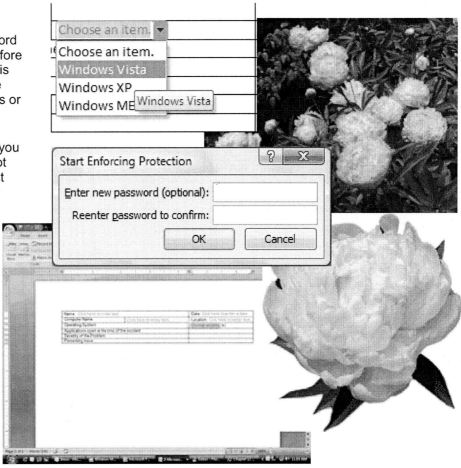

Microsoft Word 2007 Exam 77-601 Topic: 6. Sharing and Securing Content
6.2. Control document access
6.2.3. Set passwords

Word: Who Done It? Page 1 2 3 4 5 6 7 8 9 10 11 12 13 14 15 16 17 18 19 20 21 22 23 More

Developer -> Combo Box Control

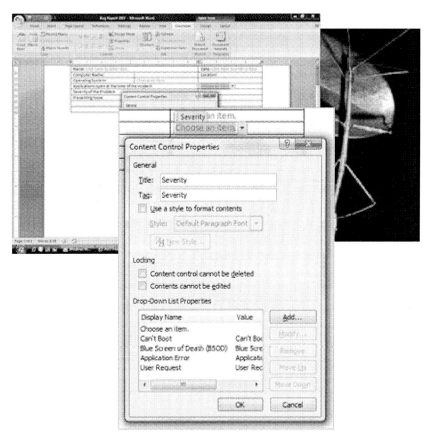

Add Another Combo Box
Go back to the **Developer** tab and **unprotect** the document so that you can add two more Combo Box Controls to the form.

Try it: Make an Application Combo
Place your cursor in Cell B4
Go to the **Developer** tab
Select **Combo Box** from the Controls
Go to the **Properties**
Title: Applications
Drop-Down List:
Word
Excel
Outlook
PowerPoint
Access

Make a Severity Combo
Title: Severity
Drop-Down List:
Can't Boot
Blue Screen of Death (BSOD)
Application Error
User Request

Word: Who Done It? Page 1 2 3 4 5 6 7 8 9 10 11 12 13 14 15 16 17 18 19 20 21 22 23 More

Insert -> WordArt

Insert WordArt
Microsoft Office has an option for creating banners with 3Dimensional letters and vibrant color gradients.

Try it: Create a banner
Go to the **Insert** tab
Click **WordArt** and select a **Style**

You will be prompted by a screen that says: YOUR TEXT HERE

Type: Bug Report.

What Do You See? By default, the text is 36 pt and the Font is Impact. You can edit the formatting if you wish.

Microsoft Word 2007 Exam 77-601 Topic: 3. Working with Visual Content
3.3. Format text graphically
3.3.1. Insert and modify WordArt

Word: Who Done It? Page 1 2 3 4 5 6 7 8 9 10 11 12 13 14 15 16 17 18 19 20 21 22 23 More

WordArt Tools -> Format

WordArt Tools

Some objects have special Ribbons that show up when you need them.

When you Insert WordArt, you will see a new **WordArt Tools** Ribbon. This is a big box of crayons that you can try. <grin>

Try it: Format the WordArt
First, select the WordArt
Go to **WordArt Tools ->Format**
Change the **Text Wrapping** to Tight
Play with the Styles, Shadows and 3-D

Move the WordArt above the Table

Memo to Self: Sometimes it is difficult to place anything above the Table.

If the WordArt ends up BEHIND the table, go to UNDO and try this. Place your cursor in Cell A1 in the table. Hit the Enter key on your keyboard to create a blank line. Try to move the WordArt, now.

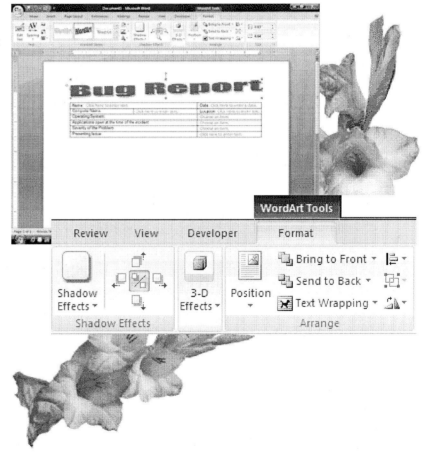

Microsoft Word 2007 Exam 77-601 Topic: 3. Working with Visual Content
3.3. Format text graphically
3.3.1. Insert and modify WordArt

Word: Who Done It? Page 1 2 3 4 5 6 7 8 9 10 11 12 13 14 15 16 17 18 19 20 21 22 23 More

Home -> Paragraph ->Line Spacing

Format for Printing

Up to this point, you have been designing a form that can be completed on the computer.

What changes should you consider if this form will be printed and handed out?

Your form has to have enough space for people to print the information by hand. If you increase the size of the text, then there will be more space between the lines. 14 pt is the minimal size for folks to write in. You can also increase the **Line Spacing**.

Try it: Format the Text
Click on the Table
Go to **Table->Layout ->Select Table**
Go back **Home**
Find the **Paragraph** group
Choose a **Line Spacing**

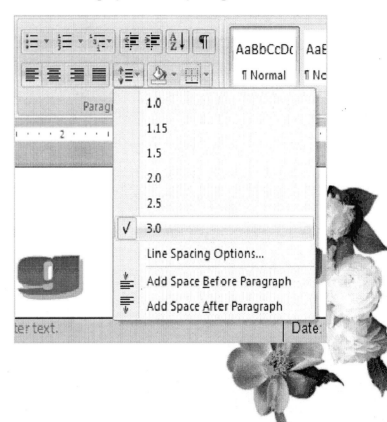

Microsoft Word 2007 Exam 77-601 Topic: 2. Formatting Content
2.1. Format text and paragraphs
2.1.4. Format paragraphs: Change line spacing

Word: Who Done It? Page 1 2 3 4 5 6 7 8 9 10 11 12 13 14 15 16 17 18 19 20 21 22 23 More

Table Tools -> Design -> Table Style

Finish the Form
Time to take off the training wheels, lock the form, and hand it out. The form was developed with a Table. Finish the form by formatting the Borders and Shading.

Try it: Format the Table
Click once on the Table to select it.
Go to the **Table Tools**.
Find the **Design** tab.

There are two sets of options you can choose to make your information easy to read: Use the **Table Style Options** to format the Header Row, Banded Rows, and First Column if you wish.

Table Styles offer Quick Style formatting for coloring the borders and shading in this Bug Report form.

What is the purpose of a Bug Report?
Where the Computer Mama works the saying goes: If it ain't written down, it didn't happen.

Word: Who Done It? Page 1 2 3 4 5 6 7 8 9 10 11 12 13 14 15 16 17 18 19 20 21 22 23 More

Page Layout -> Themes

Apply Themes

The Quick Style table formatting in the Table Design Tools identified special aspects of this table: Header Row, First Column, etc. Microsoft Office 2007 has another set of formatting tools on the **Page Layout** Ribbon: **Themes**.

Themes are applied to the entire Word document, all of the pages. The Themes include **Color** palettes, **Font** libraries, and graphic **Effects** templates.

Try it: Apply Themes
Go to the **Page Layout** Ribbon.
Select **Themes: Built In.**
Choose a Theme from the library.

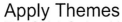

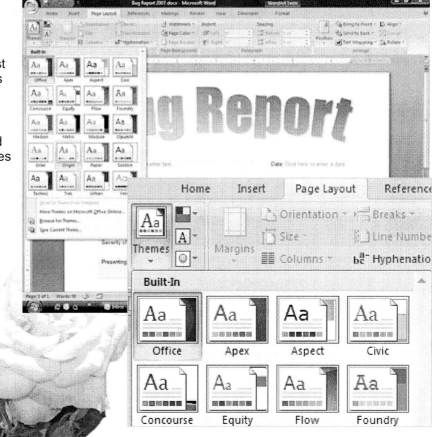

Microsoft Word 2007 Exam 77-601 Topic: 1. Creating and Customizing Documents
1.1.3. Format documents using themes: Apply themes

Word: Who Done It? Page 1 2 3 4 5 6 7 8 9 10 11 12 13 14 15 16 17 18 19 20 21 22 23 More

Page Layout -> Themes

Apply Color Themes

There are about 21 Built In color Themes. The color chips, from left to right, format the Font, Background, and Accents.

Try it: Use Theme Colors
Go to **Page Layout -> Themes**.
Select a **Built In** Theme.

What Do You See? At the bottom of the Built In Colors is the option to **Create New Theme Colors**.

Keep going...

Microsoft Word 2007 Exam 77-601 Topic: 1. Creating and Customizing Documents
1.1.3. Format documents using themes: Apply themes

Word: Who Done It? Page 1 2 3 4 5 6 7 8 9 10 11 12 13 14 15 16 17 18 19 20 21 22 23 More

Page Layout -> Themes ->Create New Theme Colors

Modify the Theme Color
Try it: Create New Theme Colors
Go to **Page Layout -> Themes.**
Go to **Create New Theme Colors.**
Edit the **Theme colors.**

The Theme colors are applied to the Text and Background in your document. The Accent colors are applied to all of the Shapes, SmartArt and Graphs.

Memo to Self: Hyperlinks change color depending on if you have not seen the link (blue) or if you have already followed that link (purple.) You may want to leave the default colors. It's what people are expecting.

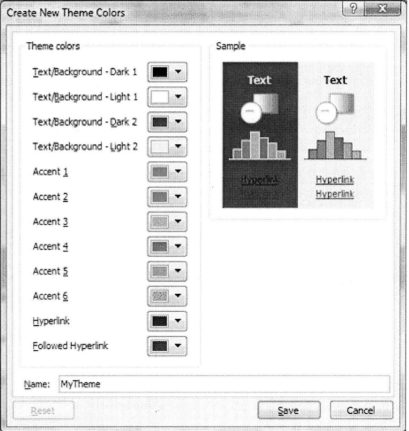

Microsoft Word 2007 Exam 77-601 Topic: 1. Creating and Customizing Documents
1.1.4. Customize themes: Colors

Word: Who Done It? Page 1 2 3 4 5 6 7 8 9 10 11 12 13 14 15 16 17 18 19 20 21 22 23 More

Page Layout -> Themes ->Create New Theme Fonts

Modify Theme Fonts
The Font libraries use different type faces for the headlines and body type. Keep in mind that headlines can be fanciful and creative. Body text, or Normal, should be Sans Serif and easy to read.

Try it: Create New Theme Colors
Go to **Page Layout -> Themes**.
Go to **Create New Theme Fonts**.
Edit the **Theme Fonts.**

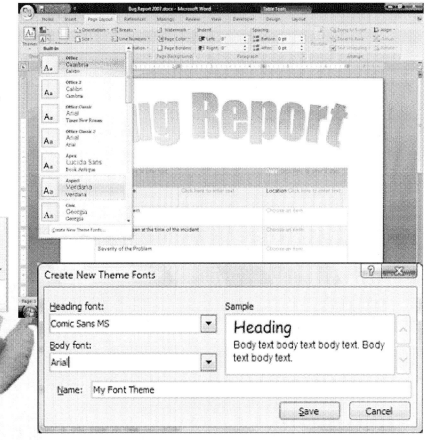

Microsoft Word 2007 Exam 77-601 Topic: 1. Creating and Customizing Documents
1.1.4. Customize themes: Fonts

Word: Who Done It? Page 1 2 3 4 5 6 7 8 9 10 11 12 13 14 15 16 17 18 19 20 21 22 23 More

Page Layout -> Themes ->Theme Effects

Modify Theme Effects
Effects format the appearance of the Graphs and SmartArt that you use to communicate your message. These tools are just too cool. <grin>

Before You Begin: Create the SmartArt
Go to **Insert ->SmartArt**.
Select the **Process** category. Choose an arrow and enter the following type: Identify, Report and Repair.

Now Try it: Create New Theme Effects
Go to **Page Layout -> Themes**.
Apply a **Theme Effect**.

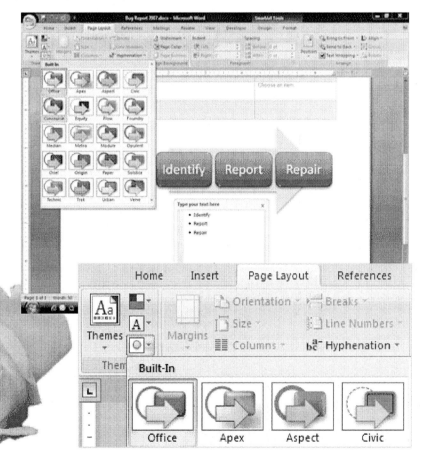

Microsoft Word 2007 Exam 77-601 Topic: 1. Creating and Customizing Documents
1.1.4. Customize themes: Effects

Word: Who Done It? Page 1 2 3 4 5 6 7 8 9 10 11 12 13 14 15 16 17 18 19 20 21 22 23 More

Page Layout -> Themes ->Theme Effects

Set Theme as Default

Say you spent some time designing your own Theme. You selected your Color palette, Fonts and Effects. You can save your Theme as the **Default** theme used with each new document that you create.

Try it: Set Theme as Default
Go to **Page Layout -> Themes.**
Go to **Save Current Themes.**
You will be prompted to enter a name for your new Office theme (*.thmx).

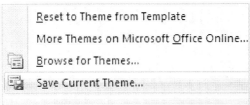

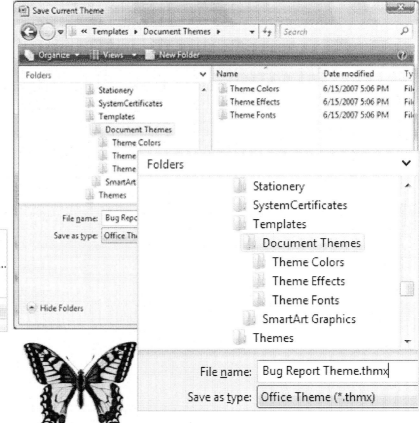

Memo to Self: You can also use this menu to Restore the Theme to one of the built in templates.

Microsoft Word 2007 Exam 77-601 Topic: 1. Creating and Customizing Documents
1.1.3. Format documents using themes: Set themes as default

Word: Who Done It? (Page 1) 24 25 26 27 28 29 30 31 32 33 34 35 36 37 38 39 40 41 42 43 44 45 46 47 48 49

Office -> Prepare ->Inspect Document

Prepare to Share

This form looks good. You need to send this around the office and get it reviewed and proofed. Before you distribute this form on the departmental server or company website, you should prepare it for publication.

Try it: Inspect the Document
Go to **Office ->Prepare**
Select **Inspect Document**

Memo to Self:
Go to Developer ->Protect Document and **Stop Protecting** the document before preparing for distribution.

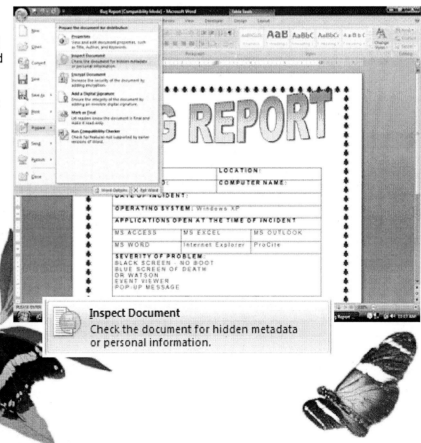

Microsoft Word 2007 Exam 77-601 Topic: 6. Sharing and Securing Content
6.1. Prepare documents for sharing
6.1.3. Remove inappropriate or private information using Document Inspector

Word: Who Done It? (Page 1) 24 25 26 27 28 29 30 31 32 33 34 35 36 37 38 39 40 41 42 43 44 45 46 47 48 49

Office -> Prepare ->Inspect Document

Document Inspector
Your Microsoft Word file contains;
Document Properties
Comments
XML Data
perhaps even Hidden Text...

Click on Inspect
Microsoft Word will alert you if anything is discovered. You can choose to accept or cancel the options.

You do not have to strip a document of all of the information. For example, many firms use the **Document Properties** to identify Title, Author, and Keywords for search and archive.

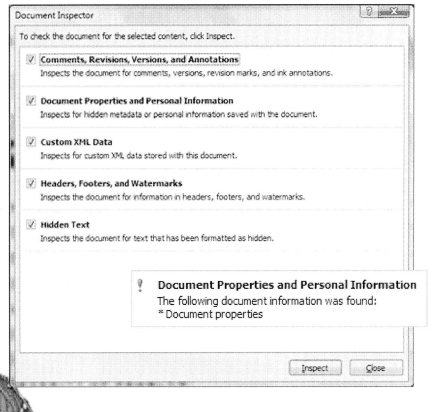

Microsoft Word 2007 Exam 77-601 Topic: 6. Sharing and Securing Content
6.1. Prepare documents for sharing
6.1.3. Remove inappropriate or private information using Document Inspector

Word: Who Done It? (Page 1) 24 25 26 27 28 29 30 31 32 33 34 35 36 37 38 39 40 41 42 43 44 45 46 47 48 49

Office -> Prepare ->Encrypt Document

Password Protect the File
You can protect your document with a password. A strong password includes letters and numbers. It should not be your name.

Try it: Encrypt the Document
Go to **Office ->Prepare.**
Select **Encrypt Document.**
Type a Password.

Security is a genuine concern. Hence, there is no "Undo" if you forget the password. Neither is there a tool that will let you remove the encryption.

Please write down your passwords!

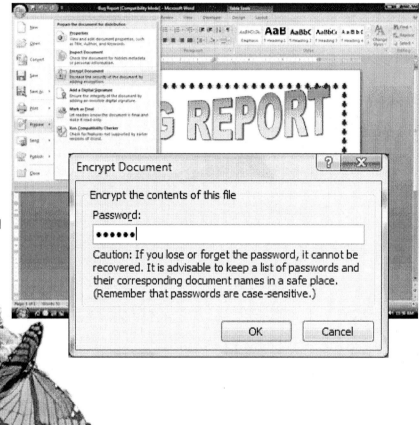

Microsoft Word 2007 Exam 77-601 Topic: 6. Sharing and Securing Content
6.2. Control document access
6.2.3. Set passwords

Word: Who Done It? (Page 1) 24 25 26 27 28 29 30 31 32 33 34 35 36 37 38 39 40 41 42 43 44 45 46 47 48 49

Office -> Prepare -> Add a Digital Signature

Attach a Digital Signature

A **Digital Signature** is similar to signing your paper forms in ink. Please note: the evidentiary laws on Digital Signatures vary. Microsoft cannot warrant the legality in every instance. That said, the purpose of the Digital Signature is to validate the integrity of your file.

The validation begins by identifying yourself with a Digital ID, which you can set up through Microsoft, or through another firm online.

Microsoft Word 2007 Exam 77-601 Topic: 6. Sharing and Securing Content
6.3. Attach digital signatures

Word: Who Done It? (Page 1) 24 25 26 27 28 29 30 31 32 33 34 35 36 37 38 39 40 41 42 43 44 45 46 47 48 49

Review -> Track Changes

Track Changes

Each person that edits this document can leave their finger prints behind. Microsoft Word can track the changes and document who made them.

Try it: Track Changes
Go to **Review ->Track Changes.**

When you edit the words, graphics, or formatting, you will see the Markup in red. The example on this page indicates that eBeth, on December 18, 2007, inserted the word: User.

Memo to Self: When you click on the **Track Changes** button, you enable the tracking. You can disable the tracking when you click **Track Changes**, again.

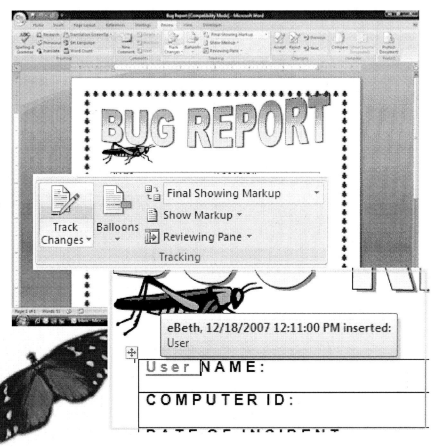

Word: Who Done It? (Page 1) 24 25 26 27 28 **29** 30 31 32 33 34 35 36 37 38 39 40 41 42 43 44 45 46 47 48 49

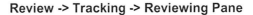

Manage Tracked Changes

A big proposal or report may have many sections and several editors. Microsoft Word has a new method for navigating the tracked changes.

Try it: Open the Reviewing Pane
Go to **Review->Tracking**.
Select **Reviewing Pane**.

You should see a new navigation map, the **Reviewing Pane**, on the left side. The Reviewing Pane shows the revisions, organized by location.

You can **Accept or Reject** the changes.

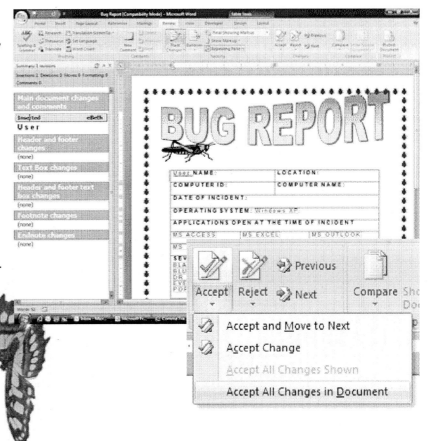

Microsoft Word 2007 Exam 77-601 Topic: 5. Reviewing Documents
5.3. Manage track changes: Accept and reject changes

Word: Who Done It? (Page 1) 24 25 26 27 28 29 30 31 32 33 34 35 36 37 38 39 40 41 42 43 44 45 46 47 48 49

Review -> Tracking -> Show Markup

Display Markup
You can determine how much you wish to see in the Reviewer's pane on the left side of your tracked document.

Try it: Show the Markup
Go to **Review -> Tracking**.
Select **Show Markup**.

What Do You See? The Markup menu has a check mark for each topic that is displayed: Comments, Ink, Insertions and Deletions and Formatting,

There is also an option for displaying the Markup Area with a highlight.

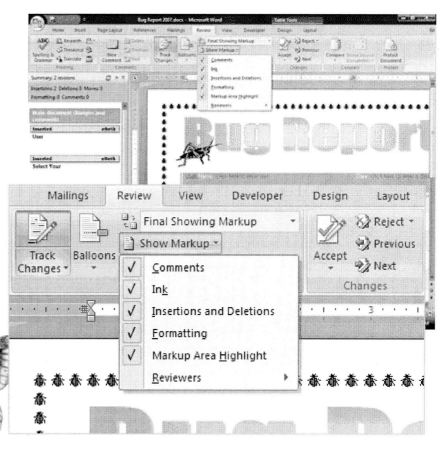

Microsoft Word 2007 Exam 77-601 Topic: 5. Reviewing Documents
5.3. Manage track changes
5.3.1. Display markup

Word: Who Done It? (Page 1) 24 25 26 27 28 29 30 31 32 33 34 35 36 37 38 39 40 41 42 43 44 45 46 47 48 49

Show Tracked Changes by Reviewer

Each person that makes changes to your document is identified. You can show the changes for **All Reviewers**, or a single reviewer.

Try it: Show the Markup
Go to **Review -> Tracking**.
Select **Show Markup**.
Go to **Reviewers**.

Did You Notice? The reviewers are assigned their own color in the document markup.

Review -> Tracking -> Show Markup -> Reviewers

Microsoft Word 2007 Exam 77-601 Topic: 5. Reviewing Documents
5.3. Manage track changes
5.3.1. Display markup: Display tracked changes and comments by reviewer

Word: Who Done It? (Page 1) 24 25 26 27 28 29 30 31 32 33 34 35 36 37 38 39 40 41 42 43 44 45 46 47 48 49

Review -> Tracking -> Balloons

Use Tracking Balloons
In the previous example, the document was displayed as Final Showing Markup.
Here is a different view of the tracking.

Try it: Use Tracking Balloons
Go to **Review -> Tracking**.
Turn off the **Reviewing Pane**.
Select **Balloons**.

What Do You See? When you click on **Show Revisions in Balloons** you should see a new, grey margin on the right side of your document.

The example on this page shows that new text was inserted.

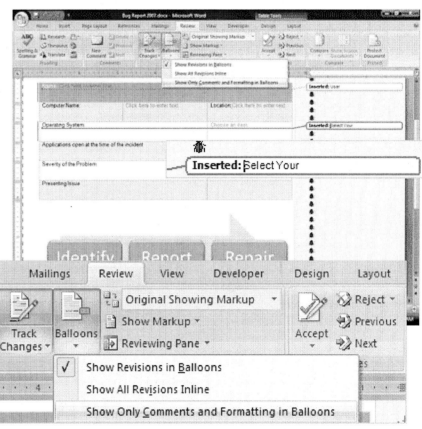

Microsoft Word 2007 Exam 77-601 Topic: 5. Reviewing Documents
5.3. Manage track changes
5.3.3. Change tracking options

Word: Who Done It? (Page 1) 24 25 26 27 28 29 30 31 32 33 34 35 36 37 38 39 40 41 42 43 44 45 46 47 48 49

Review -> Changes -> Accept

Modify Insertions and Deletions Effectively

By default, all of the changes are tracked. You may see balloons for text that was inserted, deleted or formatted. If you run your mouse over the tracked changes you should also see who made this correction and when it was made.

Try it: Modify the Changes
You can right-mouse click to **Accept** or **Reject** the tracked changes. The Computer Mama noted that you can also edit the text in the balloon if you wish.

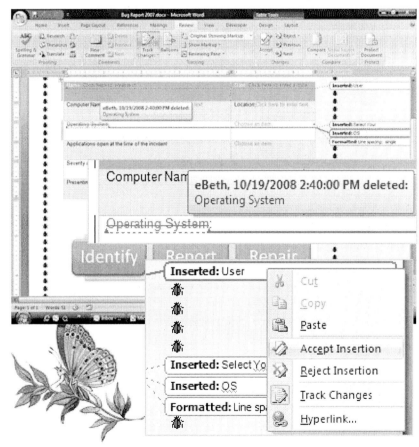

Microsoft Word 2007 Exam 77-601 Topic: 5. Reviewing Documents
5.3. Manage track changes
5.3.3. Change tracking options: Modify insertions and deletions

Word: Who Done It? (Page 1) 24 25 26 27 28 29 30 31 32 33 34 35 36 37 38 39 40 41 42 43 44 45 46 47 48 49

Review -> Tracking -> Change Tracking Options

Change Tracking Options

Microsoft Word 2007 lets you select how you would like to track and display the changes. You can even select which colors you would like.

Try it: Change the Tracking Options
Go to **Review -> Tracking**.
Select **Change Tracking Options**.

What Do You See? At the top of the list are the **Markup** options. By default, the color is assigned by **Author**. Anything that is deleted is formatted with a strikethrough ~~like this~~. Again, each author gets their own color.

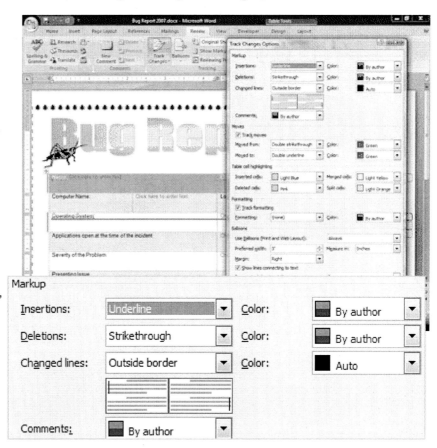

Microsoft Word 2007 Exam 77-601 Topic: 5. Reviewing Documents
5.3. Manage track changes
5.3.3. Change tracking options

Word: Who Done It? (Page 1) 24 25 26 27 28 29 30 31 32 33 34 35 36 37 38 39 40 41 42 43 44 45 46 47 48 49

Review -> Tracking -> Track Change Options

Track Changes and Moves
What Else Do You See?
You can also track if the text was **Moved**. There are two tracking marks when you move text. The text that you moved is left in it's original place, but it will have a Double strikethrough. When you paste the text into a new place in your document, the text will be tracked with a Double underline.

Tracking Your Font Challenged Colleagues. You can also mark if your artistically inclined reviewers have changed the formatting. Each reviewers' suggestions will be tracked with a different color.

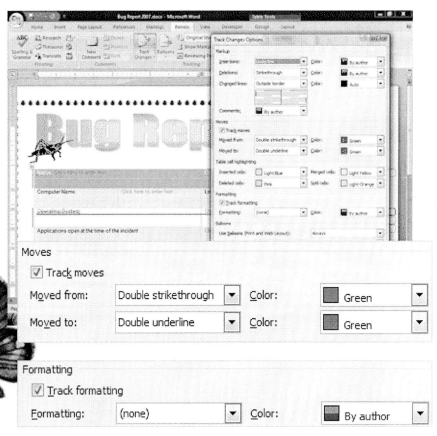

Word: Who Done It? (Page 1) 24 25 26 27 28 29 30 31 32 33 34 35 36 37 38 39 40 41 42 43 44 45 46 47 48 49

Review -> Compare

Compare Documents

When many people work together, there may be problems with different versions being edited in different offices. How do you compare the documents or combine the work?

Create Two Similar Documents
This lesson works if you have two versions of the same document. Say you opened the Bug Report that we have been editing. Make a new version:
Go to **File-> Save As**
Type a new file name.
Change some of the labels in the second version of the Bug Report.

Try it: Compare two documents
Go to **Review ->Compare.**

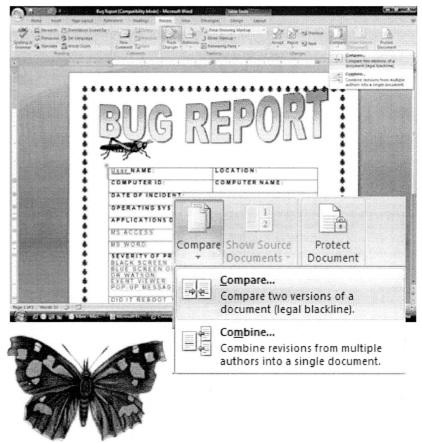

Word: Who Done It? (Page 1) 24 25 26 27 28 29 30 31 32 33 34 35 36 37 38 39 40 41 42 43 44 45 46 47 48 49

Review -> Compare

Compare and Merge
The process begins by selecting the two documents that you want to compare. Click on the yellow folder icon to browse and open the similar documents.

The original document is on the left. The revised document is on the right. You can label the changes with the name of the author if you wish.

Comparison Settings
You can Show or Hide the **Comparison settings** by clicking on the button that says: **More or Less**.

Keep going...

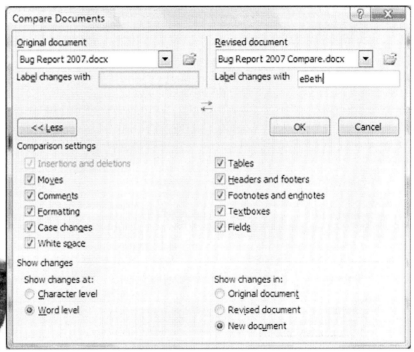

Word: Who Done It? (Page 1) 24 25 26 27 28 29 30 31 32 33 34 35 36 37 38 39 40 41 42 43 44 45 46 47 48 49

Review -> Compare

Comparison Settings
This process will compare the insertions, deletions, moves, comments, formatting, tables, Headers and Footers. This seems to be very comprehensive.

What Do You Want to Do? You have the option to show the changes in the **Original** document, a **Revised** document or a **New** document.

In this example, please choose to create a **New** document.

Keep going...

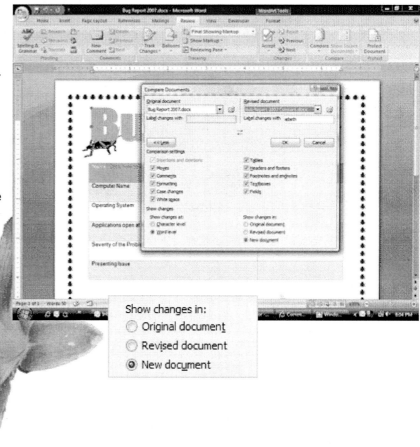

Word: Who Done It? (Page 1) 24 25 26 27 28 29 30 31 32 33 34 35 36 37 38 39 40 41 42 43 44 45 46 47 48 49

Review -> Compare -> Show Source Documents

Compare Results

The new document will summarize the revisions. At first, the screen may appear very busy. On the left you should see the **Reviewing** pane. On the right you may see the original document as well as the revised document in little windows.

You can Show or Hide the **Source Documents** as you compare the original and the changes. The options include:
Hide Source Document
Show Original
Show Revised and
Show Both.

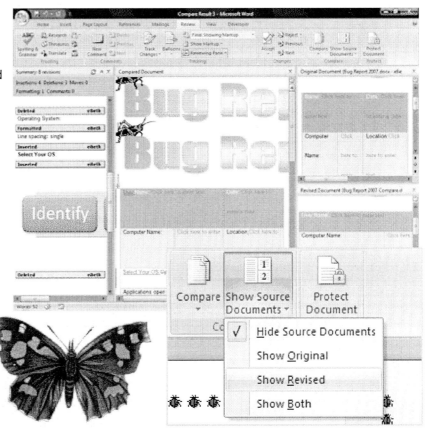

Microsoft Word 2007 Exam 77-601 Topic: 5. Reviewing Documents
5.2. Compare and merge document versions
5.2.1. Compare document versions: Manage multiple documents simultaneously

Word: Who Done It? (Page 1) 24 25 26 27 28 29 30 31 32 33 34 35 36 37 38 39 40 41 42 43 44 45 46 47 48 49

Review -> Compare -> Show Source Documents

Combine Revisions

You can use the **Reviewing** pane to **Accept** or **Reject** the tracking changes from several authors or editors.

Each revision indicates who (author) did (46 revisions) what (tracking details). The final document combines the work of many authors into one file.

Very good.

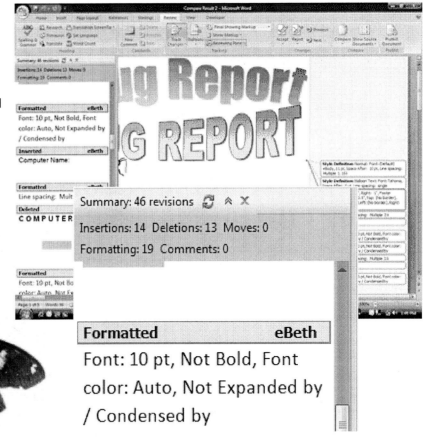

Microsoft Word 2007 Exam 77-601 Topic: 5. Reviewing Documents
5.2. Compare and merge document versions
5.2.3. Combine revisions from multiple authors

Word: Who Done It? (Page 1) 24 25 26 27 28 29 30 31 32 33 34 35 36 37 38 39 40 41 42 43 44 45 46 47 48 49

Review -> Comments

Add Comments
Comments are another method for adding your observations to the review.

Try it: Add a Comment
Place your cursor somewhere you would like to insert a comment.
Go to **Review ->Comments.**

You should see a new balloon on the right side of the document. Type some sample text. See if you can edit and modify the comments.

Microsoft Word 2007 Exam 77-601 Topic: 5. Reviewing Documents
5.4. Insert, modify, and delete comments

Word: Who Done It? (Page 1) 24 25 26 27 28 29 30 31 32 33 34 35 36 37 38 39 40 41 42 43 44 45 46 47 48 49

Review -> Proofing -> Research

Better Look It Up
Say you needed to confirm something for this document. You can use the **Proofing** tools to **Research** the spelling as well as the meaning. You can even look up similar words in the Thesaurus. Very nice.

Try it: Research a Word
Go to **Review -> Proofing**.
Select **Research**.

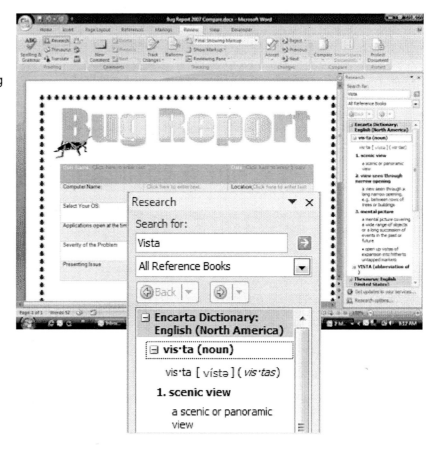

Microsoft Word 2007 Exam 77-601 Topic: 1. Creating and Customizing Documents
1.4. Personalize Office Word 2007
1.4.2. Change research options

Word: Who Done It? (Page 1) 24 25 26 27 28 29 30 31 32 33 34 35 36 37 38 39 40 41 42 43 44 45 46 47 48 49

Review -> Proofing ->Research

Research Options
Microsoft Office 2007 uses the Internet to provide fast, up to date information. You can choose which sources you would prefer to use in your research,

Try it: Edit the Research Options
Go to **Review -> Proofing.**
Select **Research.**
Go to **Research Options** at the bottom of the blue task pane.

What Do You See? The Options include many online Research sites such as the **Encarta** Encyclopedia. The **MSN Money** Stock Quotes are also included. You can add other sites from different countries if you wish.

Microsoft Word 2007 Exam 77-601 Topic: 1. Creating and Customizing Documents
1.4. Personalize Office Word 2007
1.4.2. Change research options

Word: Who Done It? (Page 1) 24 25 26 27 28 29 30 31 32 33 34 35 36 37 38 39 40 41 42 43 44 45 46 47 48 49

Office -> Prepare ->Mark As Final

Make it Final!

As Arlo Guthrie used to sing, this Word file has been "inspected, injected, rejected, dejected and selected." Here are the steps to mark the document as **Final**.

Try it: Mark the Document Final
Go to **Office ->Prepare**.
Select **Mark as Final**.

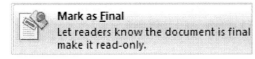

Memo to Self: When a document is marked final it will be **Read Only**. That means the editing, formatting and reviewing commands will not be available. You should see a small icon in the bottom left corner that indicates **Marked as Final**.

Microsoft Word 2007 Exam 77-601 Topic: 6. Sharing and Securing Content
6.2. Control document access
6.2.2. Mark documents as final

Word: Who Done It? (Page 1) 24 25 26 27 28 29 30 31 32 33 34 35 36 37 38 39 40 41 42 43 44 45 46 47 48 49

Customize Word 2007

The only thing left to learn in Microsoft Word 2007 is how to personalize your Office: toolbars, research options, initials and even the view. Starting at the top:

Try it: Customize the Quick Access
Go to the **Quick Access toolbar.**
Select **Customize....**

What Do You See? The options that are selected have a check mark. The short list includes the common commands: new, open, save, print and E-mail.

You can go to **More Commands** to choose another function if you wish

Customize the Quick Access Toolbar

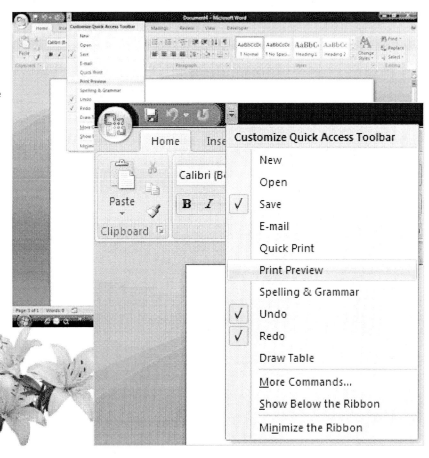

Microsoft Word 2007 Exam 77-601 Topic: 1. Creating and Customizing Documents
1.4. Personalize Office Word 2007
1.4.1. Customize Word options: Customize the Quick Access Toolbar

Word: Who Done It? (Page 1) 24 25 26 27 28 29 30 31 32 33 34 35 36 37 38 39 40 41 42 43 44 45 46 47 48 49

Office -> Prepare ->Properties

Personalize Word 2007
When Microsoft Office is installed the program prompts you to enter your name, initials, and the name of your organization. This information is saved in the document properties. Here are the steps to edit that data.

Try it: Edit the Properties
Go to **Office ->Prepare**.
Select **Properties**.

What Do You See?
The Document Properties include:
Author
Title
Subject.

You can add:
Keywords
Category (remember Outlook?)
Status

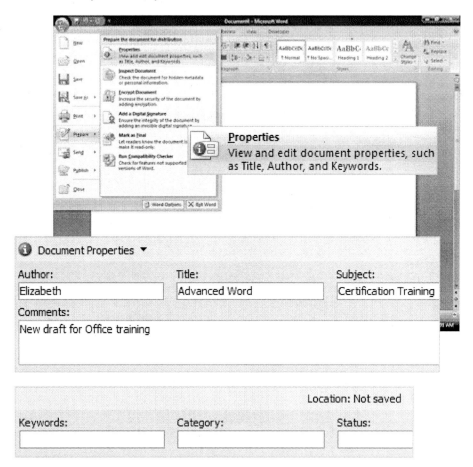

Word: Who Done It? (Page 1) 24 25 26 27 28 29 30 31 32 33 34 35 36 37 38 39 40 41 42 43 44 45 46 47 48 49

Office -> Word Options ->Popular

Popular Word Options
Almost every part of Word can be edited to fit. You can use the Word Options to turn the Developer toolbar on or off, change the Display, as well as tweak the Proofing tools.

The **Word Options** can be found under the Office logo. This discussion will start with the **Popular** choices.

Try it: Edit the Word Options
Go to **Office ->Word Options.**
Go to the **Popular options.**

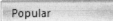

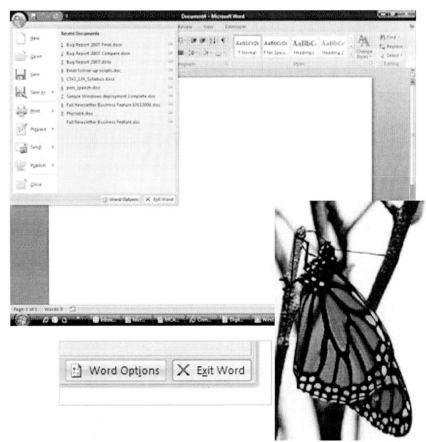

Microsoft Word 2007 Exam 77-601 Topic: 1. Creating and CUSTOMIZING Documents
1.4. Personalize Office Word 2007
1.4.1. Customize Word options

Word: Who Done It? (Page 1) 24 25 26 27 28 29 30 31 32 33 34 35 36 37 38 39 40 41 42 43 44 45 46 47 48 49

Office -> Word Options ->Popular

Popular Word Options

What Do You See? The Popular tab includes the top selections:
Show the Mini Toolbar.
Enable Live Preview.
Show the Developer toolbar.

Did You Notice? This is where you can set your User name and initials. Anything you type here will be added automatically to the Properties for all of the new Word documents you create.

Memo to Self: By default, Microsoft Word 2007 opens your E-mail attachments in the **Full Screen Reading View**. You can disable this feature if you wish: just uncheck this option.

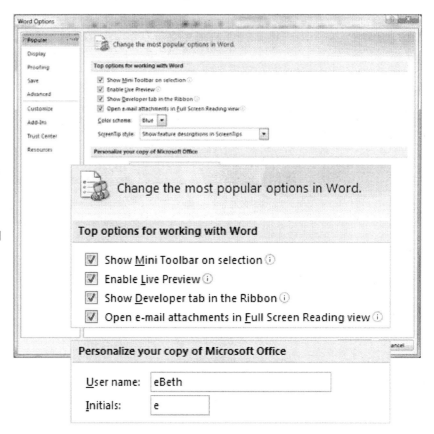

Microsoft Word 2007 Exam 77-601 Topic: 1. Creating and Customizing Documents
1.4. Personalize Office Word 2007
1.4.1. Customize Word options: Disable the open e-mail attachments feature in reading mode

DONE

Word: Who Done It? (Page 1) 24 25 26 27 28 29 30 31 32 33 34 35 36 37 38 39 40 41 42 43 44 45 46 47 48 49

Office -> Cookies

Who Done It?
This lesson began with a blank page in Microsoft Word and finished with a form that enables your users to fill in the form with a drop-down box.

The document has been prepared and inspected for distribution. There is a process in place for tracking comments, changes, and security.

Who done it? You done it: you mastered Microsoft Word.

You can have all of the cookies.

Test Yourself

1. Which parts of a Theme can be customized?
(Select all correct answers)
a. Colors
b. Fonts
c. Effects (of Graphs and Smart Art)
d. None of the above
   Tip: Advanced Guide to Word, p 107, 108, 109, 110

2. Which of the following tools are used for checking for private information in the properties of a Word document?
a. Property Inspector
b. Document Inspector
c. Document Checker
d. Personal Information Checker
   Tip: Advanced Guide to Word, p 112

3. To turn on Track Changes, click the Track Changes button.
a. TRUE
b. FALSE
   Tip: Advanced Guide to Word, p116

4. Which of the following are true about Track Changes?
(Select all correct answers)
a. You can view the changes in a Reviewing Pane
b. Changes can be accepted or rejected
c. Comments are a part of Track Changes
d. Each Reviewer is assigned their own color of Mark Up
   Tip: Advanced Guide to Word, p 117

5. Which of the following are parts of the Mark Up?
(Select all correct answers)
a. Comments
b. Ink
c. Insertions
d. Deletions
e. Formatting
   Tip: Advanced Guide to Word, p118

6. Compare and Merge combines two documents and marks difference between them.
a. TRUE
b. FALSE
   Tip: Advanced Guide to Word, p 125

12. What is the command to change the Reference Options?
a. Reference-> Reference Options
b. Review-> Proofing-> Reference Options
c. Home-> Spelling and Grammar-> Reference Options
   Tip: Advanced Guide to Word, p 130, 131

13. A document that is marked Final is set to Read Only
a. TRUE
b. FALSE
   Tip: Advanced Guide to Word, p132

Downloads
2007 Bug Report Form Complete

Assessment
Word Practice Activities
Complete the **Online Quiz** for this level.

Advanced Word Skills Test
Download the instructions. Submit the Skills Test to your instructor

Self-Assessment

Skill Level-Beginning	Mastered	Needs Work	Required for my job
Create a new document			
Select, copy and paste text			
Format text			
Format columns			
Format borders and shading			
Spell and Grammar Check			
Insert a picture from ClipArt or a file			

Beginning Word is recommended if you selected "needs work" on three or more skills

Skill Level-Intermediate	Mastered	Needs Work	Required for my job
Create a watermark			
Use Headers and Footers			
Create a template			
Create a table			
Add, delete and modify rows, column and cells			
Create a Mail Merge			
Insert a bookmark or a hyperlink			
Create an on-line (Web) page			

Intermediate Word is recommended if you selected "needs work" on three or more skills

Skill Level-Advanced	Mastered	Needs Work	Required for my job
Format text with Styles			
Navigate with the Document Map			
Insert captions, footnotes or endnotes			
Create a Table of Contents			
Track Changes			
Create an on-line form			
Create a Master Document			

Advanced Word is recommended if you selected "needs work" on three or more skills.

The Comma Method

Advanced Microsoft® Word: Practice 1

RUBRIC

0	3	5	8	10
Less than 25% of items completed correctly.	More than 25% of items completed correctly	More than 50% of items completed correctly	More than 75% of items completed correctly	All items completed correctly

Each step to complete is considered a single item, even if it is part of a larger string of steps.

Objectives:

The Learner will be able to:
1. **Create a simple Form in Word at least 75% of the time**
2. **Insert Form commands into a Table at least 75% of the time**
3. **Insert a Text Field into a Form at least 75% of the time**
4. **Insert a Checkbox into a Form**
5. **Insert a Date Field into a Form**
6. **Protect a Document at least 75% of the time**
7. **Test your form design and controls**

Basic Form Design

This practice exercise requires the Developer's toolbar. Here are the steps to turn it on:

Go to **Office-> Word Options->Popular**
Check: **Show Developer Tab** in the Ribbon

Use a Table to Layout the Form

Start a new Word Document
Insert a Table with 3 row2 and 3 columns
Label the Columns with the following titles:
Name, Received Book, Response

Name	Received Book	Response
		☐ Replied to email

Add the following Form Controls:

Under the Name column, insert a **Text Field**
Under the Received Book column, insert a **Date Field**
Under the Response column insert a **Checkbox** and edit the label
Protect the Document for forms

Test the Form

Fill in the Form with a name, check the box, and a date

Save your practice document and name it: Advanced Word Practice 1

Advanced Microsoft® Word: Practice 2

Objectives:
The Learner will be able to:
1. Create new Styles at least 75% of the time
2. Modify Styles at least 75% of the time
3. Apply styles to text at least 75% of the time
4. Insert breaks
5. Use continuous section breaks
6. Insert Headers and Footers that are different for different sections
7. Create a Table of Contents with 2 levels and page numbers

Styles

This practice uses Styles to format the headlines in a sample text document. In turn, the Style can be used to create a Table of Contents. The student is also prompted to practice creating Section Breaks and developing different Headers an Footers for each Section.

Open the sample Word document: Windows Deployment Text

Create the following Styles:
Heading 1: Tahoma 14pt, Bold, Purple
Heading 2: Tahoma 12pt, Bold, Purple
Heading 3: Tahoma 10pt, Bold, Automatic

Document Formatting:
Format the headlines in the document with Heading 1, 2, and 3.

Edit the Page Layout
Insert **Section breaks** for Phase I, II, and III
Choose the option to start the new section on the next page
Add a cover page for page 1
Insert a blank page after the cover page on page 1
Add page numbers to the Footer

Create a Table of Contents
Go to **References->Table of Contents**
Use 2 levels and include the page numbers on the right side

Save your practice document and name it: Advanced Word Practice 2

Name: _____
Instructor: _____
Test Score: _____

Microsoft Skills Test

Advanced Word

☐ 1. Microsoft Word Advanced: Action Step 1
   Open a new blank document.
   Type: Status Report

☐ 2. Microsoft Word Advanced: Action Step 2
   Select the text and format it with the Heading 1 Style

☐ 3. Microsoft Word Advanced: Action Step 3
   Add a blank line after the Status Report headline

☐ 4. Microsoft Word Advanced: Action Step 4
   Insert a Table with three columns and four rows

☐ 5. Microsoft Word Advanced: Action Step 5
   Add the following labels:
   Cell A1: Name
   Cell B1: Location
   Cell C1: Date

   Cell A2: Applications

☐ 6. Microsoft Word Advanced: Action Step 6
   Use the Office Word Options to show the Developer's Toolbar

☐ 7. Microsoft Word Advanced: Action Step 7
   Add a Text form field Control in Cell A1

☐ 8. Microsoft Word Advanced: Action Step 8
   Add a Drop-Down Control in Cell B1
   Include the following locations: Rec Room, Gold Room, Library, Conservatory, Hall

Microsoft Assessment Test
Intermediate Excel

☐ 9. Microsoft Word Advanced: Action Step 9
Arrange the Drop-Down list alphabetically

☐ 10. Microsoft Word Advanced: Action Step 10
Add a Date Picker Control in Cell C1

☐ 11. Microsoft Word Advanced: Action Step 11
Create a Drop-Down list in Cell A2
Include the following applications: Word, Excel, PowerPoint, Outlook, and Visio
Add Help Text to the Control: Select your application from the list
Copy and Paste this Control to Cells B2 and C2

☐ 12. Microsoft Word Advanced: Action Step 12
Format the Table: No Borders and do not show the Gridlines

☐ 13. Microsoft Word Advanced: Action Step 13
Save the file as Your Name Word Advanced Form Design.
Please submit the spreadsheet to your instructor.

Microsoft Skills Test

Name: _____

Instructor: _____

Test Score: _____

Advanced Word

1. Microsoft Word: Headers and Footers
   Check all that are true.

   ☐ A. By default, Word inserts the page numbers in the footer at the bottom of the page
   ☐ B. To add a Header, you would use View -> Header and Footer.
   ☐ C. When you insert a page number, it automatically repaginates.

2. Microsoft Word: Headers and Footers
   Check all that are true.

   ☐ A. The Header and Footer toolbar lets you switch from Header to Footer
   ☐ B. The Headers and Footers a special Text boxes at the top and bottom of a document.
   ☐ C. The Header and Foot toolbar includes AutoText.

3. Microsoft Word: Form Design
   Which steps are needed to create a text form field? Select all that are correct.

   ☐ A. Go to View ->Toolbars ->forms
   ☐ B. To insert a text box, click on the abI button
   ☐ C. Add Help Text in the Status Bar by clicking on the Add Help Text button
   ☐ D. Confirm that there is a check mark in the Fill-enabled option.
   ☐ E. Protect your document from Form changes

4. Microsoft Word: Form Design
   You can use the columns and rows in a table to make a professional form and simplify the form design.

   ☐ A. True
   ☐ B. False

5. Microsoft Word: Form Design
   If you can't see your form field, look for the shading button on the Forms toolbar. It looks like a letter "a".

   ☐ A. True
   ☐ B. False

6. Microsoft Word: Form Design
   Here are some options you can choose for your text form field. Which of these are correct?

   ☐ A. Text Formats include: Lowercase, Uppercase, First Capital and Title Case
   ☐ B. You can leave the Maximum length unlimited to let your form users type as much as they need
   ☐ C. A date box is just a Text Form Field with a different format.

Microsoft Assessment Test
Advanced Word

7. Microsoft Word: Form Design
Here are some options you can choose for your drop-down form field. Which of these are correct?

☐ A. Selecting answers from a list means the data will be consistent
☐ B. You can change the order of the names on the list. Select a name then use the arrows on the right to move the label up or down.
☐ C. Word Forms have to be Protected before they work
☐ D. Each document can be protected for forms, tracked changes and comments

8. Microsoft Word: Styles
A blank Word document comes with over a dozen Styles formatted. Which of the following Styles are included?

☐ A. Normal text
☐ B. Heading 1, 2, and 3
☐ C. Bullets and Numbers
☐ D. Footnotes and Endnotes

9. Microsoft Word: Styles
To create your own Heading Style, select the type and go to Format-> Styles. Select Heading 1 from the list of Styles and click on Modify.

☐ A. True
☐ B. False

10. Microsoft Word: Styles
Say you had a document with 20 headlines that you formatted with the Heading1 Style. If you go to Format -> Styles and modify Heading1, all 20 headlines will change automatically.

☐ A. True
☐ B. False

11. Microsoft Word: Styles
You formatted all of the headlines in your document with Styles. When you click on the Document Map you will see a new pane on the left side of the screen that lists your headlines as an Outline.

☐ A. True
☐ B. False

12. Microsoft Word: Styles
The Table of Contents, like the Document Map, is created from the headlines you formatted with Styles.

☐ A. True
☐ B. False

Microsoft Assessment Test
Advanced Word

13. Microsoft Word: Styles
The Table of Contents, like the Document Map, creates hyperlinks to the headlines you formatted with Styles.

- [ ] A. True
- [ ] B. False

14. Microsoft Word: References
There are several reps needed to create a Table of Contents. Which of the following steps are correct?

- [ ] A. Format the Headlines with Styles
- [ ] B. Format the Headlines by going to Format ->Font
- [ ] C. Go to Insert ->Page Numbers
- [ ] D. Go to Insert -> References -> Index and Tables
- [ ] E. Select the Style and number of levels from the Table tab

15. Microsoft Word: References
You can update all of the Table of Contents or just the page numbers if you right click on the Table of Contents.

- [ ] A. True
- [ ] B. False

MCAS Word Word Beginning Word Intermediate Word Advanced

Index Advanced Microsoft Word 2007 Exam 77-601 Guide

Apply Quick Styles to Documents, 44
Apply Themes, 106

Comments: Insert and Delete, 129

Convert Tables to Text, 26
Convert Text to Lists, 27
Convert Text to Tables, 18

Customize Word 2007, 131
Customize Word 2007: Disable Reading Mode for E-Mail, 136
Customize Word 2007: Personalize User Name and Initials, 134
Customize Word 2007: Quick Access Toolbar, 70

Customize Themes, 107
Customize Themes: Colors, 108
Customize Themes: Effects, 110
Customize Themes: Fonts, 109

Date and Time: Automatic, 63
Date and Time: Modify, 63
Different First Page, 65
Digital Signatures, 115

Document Inspector: Options, 113
  Remove Annotations, 113
  Remove Hidden Text, 113

Document Properties: Key Words, 134
Document Properties: Office Button, 134

Find and Go To, 87
Find and Replace: Replace All, 86
Find and Replace: Replace Text, 85
Find and Replace: Search for Text, 86

Format Characters: Clear Format, 57
Format Documents Using Themes, 111

Format Paragraphs: Indentation, 37
Format Paragraphs: Spacing, 35
Format Paragraphs: Quotes, 82

Format Text: Drop Caps, 83
Format Text: Pull Quotes, 82

Headers and Footers : Modify, 62
Headers and Footers: Page Nos, 62

Indexes: Create, Modify and Update, 73
Indexes: Mark an Entry for Indexing, 74

Insert Blank Pages or Cover Pages, 64
Lay Out Documents: Page Numbers, 67

Lists: Change Bullet Options, 28
Lists: Change Numbering, 31
Lists: Promote and Demote Items, 29
Lists: Sort Items, 33

MCAS Word Word Beginning Word Intermediate Word Advanced

 Index Advanced Microsoft Word 2007 Exam 77-601 Guide

Manage Multiple Documents, 124
Page Breaks: Insert and Delete, 58

Prepare to Share:
 Document Inspector, 112
 Mark as Final, 132
 Restrict Permissions, 99
 Set Editing Restrictions, 99
 Set Formatting Restrictions, 100
 Set Passwords, 100
 Set Passwords, 114

Quick Parts: Fields, 69

Reference Style: MLA, APA or Chicago Manual, 77
References: Bibliographies, 79
References: Citations, 75
References: Sources, 76
References: Table of Figures, 80
Research: Change Options, 130
Restore Template Themes, 111

Section Breaks, 66
Section Breaks: Delete, 68
Section Breaks: Insert, 66
Section Breaks: Headers, Footers, 67

Set Themes as Default, 111

Styles: Apply Styles, 4
Styles: Change Fonts, 50
Styles: Change Styles, 48
Styles: Create New Style Based on Existing Style, 52
Styles: Create New Style, 53
Styles: Format Body Text, 47
Styles: Format Headings, 45
Styles: Modify Styles, 48
Styles: Reveal Style Formatting, 55
Styles: The Format Painter, 56

Table of Contents: Add Text, 72
 Create and Modify, 69
 Modify Format, 71
 Update with Selected Text, 72
 Change Position, Direction of Text, 22
 Insert and Delete Rows, Columns, 19
 Modify Table Properties, 21
 Perform Calculations, 24
 Sort Contents, 25

Tabs: Clear All Tab Stops, 17
Tabs: Clear One Tab, 17
Tabs: Leaders, 17
Tabs: Set and Clear, 13

Track Changes: Accept or Reject, 117
Track Changes: Balloon Options, 120
Track Changes: By Reviewer, 119
Track Changes: Change Options, 120
Track Changes: Display Markup, 118
Track Changes: Enable or Disable, 116